Region and Nation

Region and Nation: Politics, Economics, and Society in Twentieth-Century Argentina

Edited by

James P. Brennan and Ofelia Pianetto

St. Martin's Press
New York

REGION AND NATION

 For information, address St. Martin's Press, Scholarly and Reference Division, 175 Fifth Avenue, New York, N.Y. 10010.

ISBN 0-312-23144-X

Library of Congress Cataloging-in-Publication Data to be found at the Library of Congress.

First edition: July 2000
10 9 8 7 6 5 4 3 2 1

Contents

Introduction

In the early 1970s, one of the most renowned English-speaking historians of Argentina, James R. Scobie, published what for many years remained the best general history of the country in the language, *Argentina: A City and a Nation.* Scobie's title caught an essential national paradox, noted previously by Argentine intellectuals from Domingo Faustino Sarmiento to Ezequiel Martínez Estrada, of a bifurcated historical experience, the great metropolis and capital city versus the underdeveloped and often ignored, not the least by the Argentines themselves, hinterland. Scobie was one of the first historians to heed his own advice to give greater attention to the special character and historical development of the provinces in his posthumously-published study of late-nineteenth- and early-twentieth-century urbanization.[1] Scobie's untimely death, and at a relatively young age, robbed the *argentinistas* in the English-speaking academic community of the person best able to promote the study of provincial-regional history. Its continuing underdeveloped state has yet to recover from his loss.

Indeed, the English-language historical scholarship on Argentina, despite the strength of regional and even local studies in European and North American historiography, rather curiously has practiced a centralism of its own and has tended to follow very much a Buenos Aires-focused approach. The obvious importance of the regional economies in the nineteenth century, and only the belated, definitive consolidation of political and economic power in Buenos Aires near the end of that century, meant that historians of the nineteenth century could not practice the same centralism in their scholarship that twentieth-century historians did. As a result, there is a small body of work by North American and British scholars on the provinces, though historical research is heavily weighted in favor of Buenos Aires for the nineteenth century as well.[2]

Provincial-regional history remains, even more so than that for the previous century, largely unexplored territory for historians of twentieth-century Argentina. The list of the works of the leading English-speaking historians in particular have shown a marked proclivity for either broad, national histories (themselves quite Buenos Aires-centric) or Buenos Aires-focused studies.[3] There are a number of reasons for this fixation on Buenos Aires at the expense of the rest of the country. One is strictly practical. Archives and libraries in the provinces were, until recently, in very poor shape compared to Buenos Aires, though conditions have improved greatly in a number of provinces in recent years. Foreign scholars naturally preferred to stay in Buenos Aires where archival sources and collections, though hardly in optimum conditions, were by Argentine standards good.

Another reason for the preoccupation is that, as an intellectual proposition, the fixation on Buenos Aires was and remains valid. For over a century, greater Buenos Aires has always held a third of Argentina's population, and has been the center of the nation's economic and political power. In no other Latin American country has a single urban center arguably so dominated national life. Though socially and culturally the provinces remained important and experienced interesting historical developments of their own, sometimes representing noteworthy departures from the Buenos Aires paradigm, the nation's fortunes were more often than not played out and decided in the great metropolis. So studies of Argentine politics, the labor movement, immigration, industrialization, and now women and popular culture were and will remain quite legitimately concerned with Buenos Aires.

But the time has also come to reassess an historical scholarship so dominated by Buenos Aires; and not merely for purposes of giving the provinces and regions their due, but also because a more national, integrated perspective holds out the prospect for redefining many of the essential questions and reevaluating a number of interpretations in the great issues of twentieth-century Argentine history. It is not just a question of filling in the gaps, but of recasting historical scholarship on Argentina, and in the process perhaps reaching some very different conclusions about Argentina's twentieth-century history.

Fortunately, the Argentines themselves have recognized that federalism is not only a political issue but also an intellectual one, and there is currently a renaissance of historical scholarship on twentieth-century provincial-regional history, most of it being written by historians from the provinces themselves. Some of the best historians in Argentina are now to be found in the interior of the country, and a number of research centers, Rosario and Córdoba especially, have stood out for their dynamism and creativity. The recent establishment of an important new center of research on regional history, the *Centro de Estudios Sociales Regionales* in Rosario, and its promising journal dealing with provincial questions, *Avances*, is giving a voice to historians, particularly a younger generation who are rightfully insisting on a place for their provinces in the research and writing of what should be a truly national history. A number, though by no means all of the best of these young historians, have contributed chapters to this volume.

In what directions do these historians of the provinces appear to be taking twentieth- century historical scholarship? Perhaps the best description of their work is an insistence on diversity within pervasive national contexts. The best provincial history has been that which integrates local or regional variations into broader historical processes. The history of Radicalism, immigration, agriculture and agrarian classes, industrialization and urbanization, the working class and organized labor, Peronism—in other words, the great issues of twentieth century Argentina—have had particular expressions in the provinces, but few would argue that they are utterly unique and most recognize that they are variations on a common theme. Since the late-nineteenth-century, various national projects, often in conflict with one another, have been elaborated in Buenos Aires, the liberal and the Peronist being the most important. The provinces accompanied these national projects, indeed these projects helped to incorporate the interior into the nation, and little that has happened in this vast but in many ways homogeneous country could be said to be entirely unique. Unlike much of Latin America, there are few examples of deep cultural, racial, or ethnic conflicts in modern Argentine history, not even important geographical divisions, and we are usually not in the presence of anything entirely sui generis when we study Argentina. So Argentine provincial-regional history is always sensitive to the interaction of national contexts and local conditions. Again, the unifying concept seems to be variations on a common theme, be it the history of the Church, that of Peronism and the working class, or the role of women and gender in social and political movements.

By accounting for these variations, however, it is possible in the process to so redefine established interpretations of certain issues that it will call into question a number of historical nostrums. Labor history is one area of study to which both editors have devoted a considerable amount of their attention. Our research, and that of other historians of the provincial working class, certainly calls into question the validity of looking at working-class and labor history strictly from the point of view of the Buenos Aires paradigm. It is possible, for example, to see alternately both more conservative values among workers in the interior, such as a greater religiosity and closer relationship to the Catholic Church, and a greater proclivity for militancy, even political radicalism. The recent history of the Cordoban and Rosario working classes alone, of the *Cordobazo* and *Rosariazo,* the *clasista* movements of the 1970s, indeed their histories during the democratic restoration of the last decade, certainly revealed some marked differences from the historical experience of labor and the working class in the port city, the result naturally of different processes of class formation and of local cultural contexts. Among the local variations were different perceptions among the workers of what it meant to be a Peronist, and a local elaboration of Peronist ideology by the provincial working class and even the trade union leadership. This is only one of a number of examples that demonstrate how provincial studies can not only broaden or reshape established interpretations but also, through the comparative method, bring into sharper focus the precise character of Buenos Aires's own historical development.

This volume does indeed have a decidedly revisionist agenda. Marta Bonaudo's study of politics in Santa Fe (1890–1912) calls into question the idea of politically inactive immigrants and a demobilized civil society during the heyday of *roquismo.* Joan Supplee disputes the image of a peaceful accommodation to political reform by elites and examines the entrenched opposition, including frequent resorts to violence, by the established political clans in Mendoza at the turn of the century. Similarly, Nicholas Biddle demonstrates the persistence of "political skulduggery" and conservative resistance to free and fair suffrage in Salta during the Radical governments of the 1920s. Gabriela Olivera and Marcelo Lagos criticize nostrums of Argentine economic history and the success of export-led development in their analysis of the formation of enclave economies in La Rioja and Jujuy respectively. Jane Walter reveals the force of tradition in Argentine society in the form of the Church and Catholicism and its opposition to the Peronist regime of the 1940s and 1950s, but also the deep affinities that existed in Córdoba between its local Catholic culture and Peronism's aspirations to redefine the nation. Finally, Mónica Gordillo calls into question ideas of an "integration" of the Peronist working class with the state or integrative state ideologies in her study of the social underpinnings of the militancy and even political radicalism of the Córdoban labor movement in the 1960s. Collectively, these essays make a powerful statement for incorporating regional processes and dynamics into the rewriting of Argentine history.

By shifting the focus to the interior, the historians in this volume also demonstrate how the national narratives will be reshaped in three specific ways. First, as examples of broader national processes, thus complementing and rounding-out histories told from Buenos Aires (Bonaudo and Walter). Second, as exceptions to a metropolitan history, which show how different the interior is, thus challenging Buenos Aires-centered histories as national histories (Lagos and Olivera, and to some extent Supplee). Third, as histories of how events in the provinces actually redirected and define national history, where the action shifts from Buenos Aires to the interior (partly

Supplee and Walter, more clearly Biddle and Gordillo). In some of these essays, these purposes actually overlap. Walter's essay, for example, demonstrates how Córdoba was a unique part of a general phenomenon (the unraveling of the Peronist "alliance"). But she also shows how the peculiarities of this regional effect intensify and animate the ways in which this phenomenon occurs (the increasingly bitter clash over integralist models of social organization), and thus deflects "national" processes to provincial contexts in very concrete ways. The important point is that ignoring the interior not only leaves people out of supposedly national narratives, it also misses some of the important actors and processes in these narratives.

It is perhaps appropriate at this point to make clear what is meant by provincial-regional history. The concept of a "region" as a necessary corrective to a monolithic national history was first suggested by the French "Annales" school. But as this brilliant group of historians argued, the regional perspective represents an advance in historical research only if it illuminates broader historical configurations. We need to go beyond the banal proposition that regional history is necessary because the regions are different. Thus, what is always necessary is context. In the case of twentieth-century Argentina, there seem to us to be two pervasive, identifiable periods that establish the national historical contexts in which to frame Argentine regional history. For the decades before the 1940s, that context would be the country's rapid and for the most part successful insertion into the world economy; and furthermore how this insertion manifested itself in different ways and with different social and political implications for various parts of the country. After 1940, the context is one of crisis in the established model of development and the search for a new equilibrium, a search in which Peronism and the conflicts surrounding its presence in Argentine society are central to the country's historical experience.

From the outset, it should be clear that we include most of the province of Buenos Aires in the provincial-regional camp, despite its historically close relationship with the great port and capital city of the same name. Notwithstanding its status as the most important province in the country, studies of its twentieth-century history are few. The country's most important agricultural province, Buenos Aires was characterized by traditional social relations well into the twentieth century, ones that resembled more those found in the rest of the country than in the great metropolis. The province's major urban centers of La Plata-Berisso, Bahía Blanca, and Mar del Plata also have urban and cultural dynamics of their own, and thus belong to the domain of regional history. Indeed, there is much that distinguishes Buenos Aires the province from Buenos Aires the metropolis.

In contrast, areas such as Avalleneda, Lanus, La Matanza, and Quilmes, which by political definition belong to the province of Buenos Aires but socially, culturally, and historically form part of the Buenos Aires metropolis, we consider to belong in the twentieth century, to historical-cultural processes at work in the capital city; and indeed the historiography places them there as well. The rest of the province, again, belongs to the provincial-regional camp.

We have also deliberately used the hyphenated term "provincial-regional history" since "province" is often an arbitrary concept, an administrative and political distinction rather than a historical, cultural, or sociological one. Regional economies and regional cultures preceded the formation of individual provinces and often remain more important than political boundaries.[4] In Argentina, the "interior" turns out to be many places with many kinds of social formations, some overlap well with politics, others are ecological and defy politics. There are thus several competing proposals on the regional divisions in Argentina, but one that has often been used and seems to us the most accurate historically recognizes nine discrete regions: the port city, the Littoral (the provinces of Buenos Aires and Santa Fe), Mesopotamia (Entre Ríos and Corrientes), the center (Córdoba and San Luis), the Northeast (Chaco, Formosa, and Misiones), the Northwest (Jujuy, Salta, Tucumán, Santiago del Estero, La Rioja, and Catamarca), Cuyo (Mendoza and San Juan), Comahue (Neuquén, Río Negro, and La Pampa), and the Patagonia (Chubut, Santa Cruz, and Tierra del Fuego).[5] There are equally valid historical and cultural reasons to abolish the Comahue category altogether, placing La Pampa with the Littoral provinces and Neuquén and Río Negro with the Patagonia. But the deep south's ecology, economy, and belated integration into the nation seem to set its experience apart in some important ways from the rapid economic development and settlement of provinces such as La Pampa, Neuquén, and Río Negro that, in turn, seem to warrant a separate category of their own, distinct from those of the older regions. At any rate, our point here is not to pigeonhole the provinces but to stress that ecology and history have determined pan-provincial categories that a rigorous provincial history will take into account.

It is undoubtedly true, however, that the historians of the provinces have tended to respect political boundaries and to study their individual provinces, perhaps the one glaring weakness in the recent renaissance of scholarship on regional history. This was probably a natural tendency, given its relatively undeveloped state and the practical difficulties of conducting research on regions spanning several provinces. The following chapters are thus, for the most part, all centered on individual provinces, and the historical analyses are confined to problems restricted to those provinces. We have not been able to cover the entire country. Certain provinces, indeed entire regions such as the Northeast and the Patagonia, have still barely been studied and await their historians. Thus there are no chapters to be found on those very interesting parts of the country. On the other hand, one of the more important and studied provinces, Córdoba, has two chapters, while others such as Jujuy, La Rioja, Mendoza, Salta, and Santa Fe each have one.

Nor have we been able to devote a chapter to the history of the last 25 years in this century. There is undoubtedly much that was particular to the provinces in the contemporary history of Argentina. The state terror unleashed by the military governments of the *Proceso* (1976–83) fell especially hard on the provinces, and one such as Córdoba, for reasons undoubtedly having to do with its volatile history in the 1960s and 70s, suffered a disproportionate amount of that terror. The democratic restoration of the last decade also appears to have provincial dynamics of its own, as recent social protests in Santiago del Estero, Jujuy, and Neuquén have dramatically demonstrated. It is to be hoped that future historians will pay attention to those provincial dynamics, both those historians who will study the country's black, tragic period of military rule and those who in the future may try to decipher the hopeful, conflictual, and often ambiguous process of democratic restoration and of national adjustment to a dramatically changed international economic and political context in the final years of the twentieth century. But perhaps this volume will also serve to draw attention to the need to rethink the question of periodization. The essays cover roughly a century, from the early days of democratic experimentation (Bonaudo, Supplee) to the crisis of capitalist democracy (Walter, Gordillo). A cycle seems to begin in the late-nineteenth century and end as an historical process in the 1960s and 70s. A provincial focus may help to throw the issue of historical change into sharper relief and to question the conventional temporal markers that the historiography employs.

This book is a collaboration between Argentine and North American historians of Argentina. The neglect of provincial history that is lamented at the beginning of this volume has, as discussed, shown encouraging signs in recent years of dramatically improving. Foreign historians have also begun to make interesting contributions to the new provincial-regional history. With the restoration of democracy in Argentina in 1983 and the reestablishment of politically stable conditions in the country, the small but active community of Latin Americanist professional historians started to turn its attention once again to Argentina and to return to the country and its archives. The majority of foreign scholars continued to go to Buenos Aires and work on broad national studies and historical problems rooted in the great city's conflicts and culture. But a number of them, especially younger historians from the United States working on doctoral dissertations, ventured out into the provinces to study provincial questions. A first wave of scholarship on provincial-regional history by English-speaking historians has thus started to take shape, and some of the best work to be found is by contributors to this volume.[6]

Provincial-regional history also promises to lead to a more fluid dialogue between foreign Latin American specialists and the historical profession in Argentina. In many respects, the interior of the country more closely resembles the social and cultural configurations of other Latin American societies than does cosmopolitan Buenos Aires; and the points of interest will be greater for those in the United States and elsewhere in the English-speaking world who work on the history of other Latin American societies. The formation of the Northwestern provinces' enclave economies and their social relations, which are analyzed in two of the volume's chapters, are just one example of the Argentine interior's similarities with other Latin American societies. By developing a new historiography for Argentina, historians will thus also contribute to a deeper understanding of the history of Latin America as a whole. But for the moment, we will be satisfied with other goals. The following chapters are a first attempt to offer an alternative twentieth-century history for Argentina, the history lived and experienced in the interior of the country. In order to give the history of this "other Argentina" as broad a coverage as possible, we have tried to assemble chapters on as many different provinces and as many different themes as was feasible.

The result is perhaps a rather polyphonic one. We did not choose a single theme to study in all its provincial permutations, despite the tighter unity such an approach would have given the volume. One reason for such an editorial decision is quite simply that regional history still remains so undeveloped that assembling a volume around a single theme would have meant sacrificing the quality of the scholarship for the sake of thematic unity. More importantly, it would have defeated our purpose. We hope with this book to encourage an emerging generation of Latin Americanist historians in the United States to follow the lead of historians of Brazil and Mexico and recast Argentina's history by taking into account regional dynamics. There is, however, one issue that perhaps serves, albeit obliquely, as a unifying theme. Sarmiento's old tale about Argentina as a republic of two cities, of modern, secular Buenos Aires and traditional, Catholic

Córdoba, was more than a compelling literary metaphor and expressed an historical and cultural reality in the nineteenth century. Indeed, this vision was central to James Scobie's previously cited book that tells tales of "secondary" cities of the interior as the places where modernity and tradition meet. This encounter between modernity and tradition is also present in the provinces' twentieth-century experience and all the essays evoke to some extent that tension. The essays by Lagos and Olivera on economic history are examples. Both show how a market economy adapted to, as much as it transformed, traditional agrarian and forestry cultures. This kind of confrontation in the provinces places greater friction between tradition and modernity than a purely Buenos Aires-metropolitan account, in which the forces of tradition are much more remote and therefore weaker.

Our primary objective in this volume is, however, to convey the diversity of provincial-regional contexts and to give as broad an idea as possible of the kinds of historical questions a new generation of historians is posing, and the variety of approaches they are employing to answer them. For purposes of tighter thematic organization we did largely restrict the volume to the areas that comprise the three strengths of twentieth-century Argentine historiography: political history, economic history, and social history. This book is intended to be a starting point in other ways. The editors hope the following essays will be a stimulus for future historical scholarship, suggesting new areas of research, both literally and figuratively, but also to stimulate the dialogue between Argentine historians and those in the English-speaking world in order to begin the research and writing of what should be a truly national history.

Notes

1. James R. Scobie, *Secondary Cities of Argentina. The Social History of Corrientes, Salta, and Mendoza, 1850–1910* (Stanford: Stanford University Press, 1988).
2. Among the handful of English-language studies that deal with nineteenth century provincial-regional history are: James R. Scobie, *Revolution on the Pampas. A Social History of Argentine Wheat, 1870–1910* (Austin: University of Texas Press, 1964); Donna Guy's study of elites in Tucumán, *Argentine Sugar Politics: Tucumán and the Generation of Eighty* (Tempe: Center for Latin American Studies, 1980); Mark Szchuman's study of immigration and social mobility in Córdoba, *Mobility and Integration in Urban Argentina. Córdoba in the Liberal Era* (Austin: University of Texas Press, 1988); and Miron Burgin's classic book, *The Economic Aspects of Argentine Federalism, 1820–1852* (Cambridge: Harvard University Press, 1946).

3. To mention just the works of the more prominent historians of twentieth-century Argentina: David Rock, *Politics in Argentina, 1890–1930. The Rise and Fall of Radicalism* (Cambridge: Cambridge University Press, 1975 and *Authoritarian Argentina. The Nationalist Movement, its History and Impact* (Berkeley: University of California Press, 1993); Donna Guy, *Sex and Danger in Buenos Aires. Prostitution, Family, and Nation in Argentina* (Lincoln: University of Nebraska Press, 1991); Joel Horowitz, *Argentine Unions, the State, and the Rise of Perón, 1930–1945* (Berkeley: Institute for International Studies, 1990); Richard J. Walter, *The Socialist Party of Argentina, 1890–1930* (Austin: University of Texas Press, 1977); and *Politics and Urban Growth in Buenos Aires, 1910–1942* (Cambridge: Cambridge University Press, 1993). Of these historians, Walter is the one who has most ventured into provincial history with his first book, *Student Politics in Argentina. The University Reform and Its Effects, 1918–1964* (New York: Basic Books, 1968); and *The Province of Buenos Aires and Argentine Politics, 1912–1943* (Cambridge: Cambridge University Press, 1985). Even as provocative a book as Daniel James's study of the Peronist working class, *Resistance and Integration. Peronism and the Argentine Working Class, 1946–1976* (Cambridge: Cambridge University Press, 1988) could be criticized for conflating Buenos Aires with Argentina. For example, James's very interesting discussion of popular culture, and the tango specifically as revealing working class mentalities, seems of less use in trying to decipher popular attitudes elsewhere in the country, where the tango was much less a popular musical form. The discussion of the Peronist working class generally is very influenced by James's reading of the culture of the Buenos Aires working class. The Peronist identity of workers in Córdoba, Santa Fe, Tucumán, and other parts of the country, though certainly sharing many characteristics with *porteño* workers, demonstrated a considerable degree of diversity; to be a *peronista* worker did not mean the same thing for everyone, everywhere in Argentina. James himself would probably be the first to admit this is a shortcoming, a relatively minor one, in his excellent, path-breaking book. James's current research, in collaboration with Mirta Lobato, on the Berisso meatpacking workers will undoubtedly be a major contribution to the new provincial-regional history.
4. For an excellent overview of the evolution of Argentina's various socioeconomic regions in the nineteenth century, see Jorge Luis Ossona, "La evolución de las economías regionales en el siglo XX," in *Economía e historia. Contribuciones a la historia económica argentina,* ed. Mario Rapaport (Buenos Aires: Editorial Tesis, 1988).
5. We have relied heavily on Alejandro B. Rofman and Luis Alberto Romero, *Sistema socioeconómico y estructura regional en la Argentina* (Buenos Aires: Amorrutu Editores, 1973) in working out this regional division, a regional division, it should be stressed, that is part of ongoing historical dynamic, not a frozen taxonomy, and will undoubtedly change in the future.

6. Among the dissertations written on provincial themes in the late 1980s and early 1990s are: Daniel J. Greenberg, "The Dictatorship of Chimneys: Sugar, Politics and Agrarian Unrest in Tucumán, 1914–1930," (Ph.D. diss. University of Washington, 1985); James P. Brennan, "Peronismo, Clasismo, and Labor Politics in Córdoba, 1955–1976," (Ph.D. diss. Harvard University, 1988); Joan Supplee, "Provincial Elites and the Economic Transformation of Mendoza, 1880-1914," (Ph.D. diss. University of Texas at Austin, 1988); Nicholas Biddle, "Oil and Democracy in Argentina, 1916-1930," (Ph.D. diss. Duke University, 1991).

CHAPTER ONE

Society and Politics: From Social Mobilization to Civic Participation (Santa Fe, 1890–1909)

Marta Bonaudo

The year 1890 inaugurates Argentina's modern history. The last decade of the nineteenth century unleashed changes in the economy, society, and politics that the country would struggle with well into the next century. By 1890, state power had been consolidated and a unified ruling class had put an end to chronic civil war, replacing the war lords, *caudillos,* who presided over Argentina's tempestuous fortunes in the 80 years following independence from Spain. Several decades of European immigration had also transformed the class structure, especially in the Littoral provinces of Buenos Aires and Santa Fe where the majority of the immigrants settled. A rudimentary livestock economy had been supplanted by a diverse, modern agriculture with booming export markets. The country appeared perched on an era of ever increasing prosperity and unprecedented stability of its political institutions.

But the accomplishments masked unresolved tensions in the society of this young nation. The fruits of economic growth had been monopolized by a few. Political stability had come at the expense of a participatory, popular democracy, and elite machine politics had tarnished the legitimacy of its institutions. The success of Argentina's liberal order had further concentrated economic and political power in the capital city. A province such as Santa Fe, precisely because it had participated in and experienced the bounty of the country's success more than most of the other provinces, assumed a uniquely important role as both a critic of the failures of the era and a laboratory where social experiments and political heresies thrived. The immigrants who had flocked to the province's agricultural colonies in the final decades of the nineteenth century emerged as an important voice of dissent and protest. Political experiments such as the *Unión Cívica* and its successor, the *Unión Cívica Radical*, parties demanding a genuine democratization of the economic and political sphere, emerged in this decade and with particular force in the Littoral provinces.[1] If viewed from the perspective of Santa Fe, the 1890s were a truly watershed decade, reestablishing the rules of the political game and influencing the country's political culture in myriad ways.[2]

This chapter seeks to analyze Argentine politics in the final decade of the nineteenth century and the first of the twentieth from a perspective that sheds light on the strong articulations that, in Santa Fe's public space, were woven between formal and informal ways of "doing politics."[3] The principal objective is to avoid reducing politics exclusively to electoral practices and try to recast Argentine political history by examining the different mechanisms of articulation between society and politics.[4] For some parts of the country, especially the capital city of Buenos Aires, historical studies for an earlier period have demonstrated that though informal ways of "doing politics" did not lead to the creation of a complete citizenship, they did nonetheless allow the formation of a "public space" that, in the words of Hilda Sábato, "served as a negotiating sphere between certain sectors of society and political power which allowed a considerable part of the population to get involved in public life, an involvement which had political repercussions."[5]

Sábato's study of Buenos Aires between 1850 and 1880 reveals certain similarities with the history of the areas of agricultural colonization in the central-west region of the province of Santa Fe and in Rosario, the province's largest and most important city (see map II). Nevertheless, a number of differences can be observed after 1880, the moment when there begins an attempt to unmask politics, paradoxically at the very moment when politics is being subordinated to the orderly administration of the state.[6] Beginning around 1880, the national oligarchy, known as the Generation of 1880, consolidated central political authority and incorporated most provincial elites into the national ruling party, the *Partido Autonomista Nacional* (*PAN*). But the success of the *PAN* ultimately generated an opposition. In Santa Fe, the goal of the opposition eventually came to be the eradication of elite politics and the restoration of political rights through electoral guarantees. The voices of Santa Fe's agricultural colonists (*colonos*) and of members of Rosario's small and middling bourgeoisie became one, recovering in the process an essential element of their "citizenship," a status they understood as being an "active member" of society. Slowly, the electoral system would occupy a central place in their demands, as these social actors came to understand that it constituted the key to influencing the decisions of the powerful and to establish on new foundations of legitimacy the relationship between the governing and the governed. This realization was achieved through practices that were found both within and outside the political system. The final pages of this chapter, therefore, try to reflect upon this complex transformation that takes place between the crisis of 1890 and the emergence of a regional political party of national consequence: the *Liga del Sur.*

Crisis and Revolution: Citizens and Taxpayers Mobilize

At the very moment when at the national level the revolutionaries of the *Unión Cívica Radical* (*UCR*) were proposing a qualitative change in the way of understanding and conducting politics and were debating the relationship between authority and democracy as well as the rules of succession, the *colonos* and various factions of Rosario's bourgeoisie (who share in the revolutionary effervescence unleashed by the *UCR*) were suffering a new onslaught of the order established under *roquismo*[7] of which Santa Fe's authorities were faithful executors. The 1890s began with a new tightening of control by the central political authorities, which reduced local autonomy even more. The constitutional reform of that year not only fixed the number of the agricultural settlements that had the right to acquire the status as "municipalities," thereby enhancing the power of the central authorities in the *comisión de fomento* and the *departamento,* it also made two other significant modifications in the status quo.[8]

One such modification affected the terms of representation and tarnished the regime's legitimacy by eliminating the elective character of the mayor (*intendente*), an office that henceforth was to be appointed by the provincial chief executive. The other restricted any potential electoral base, taking the vote away from foreigners. At the same time, the provincial government absorbed the powers that before belonged to local authorities: the civil registry, public education, justices of the peace, and the property registry. There is no doubt that the intent was to limit the political influence that local communities had been increasingly exercising over the course of the previous three decades, reducing them to strictly "administrative" functions.

Nevertheless, in both the center-west part of the province and in the south—especially in Rosario—previous experiences had left a deep mark. The failed attempt to consolidate the *Partido Constitucional*[9] and the successive blocking of electoral mobilization—which only the *colonos* of Esperanza seemed to have been able to surmount—did not prevent the shaping of a public space in which that formulator of public opinion, an independent press, as well as a "culture of public pressure" through mobilization, co-existed along with new forms of organization.[10]

As Tulio Halperín Donghi has noted, it is undeniable that *roquismo,* which along with its local incarnations was the very embodiment of Juan Bautista Alberdi's "possible republic," also raised issues of the "true republic," among them the problems of "politics and democracy."[11] But it is also true that cracks in the new order, challenges to it, would do little to undermine the status quo as long as prosperity existed. It is for that reason that the year 1890 has even greater significance than generally is thought, since from the socioeconomic point of view it clearly represented one of the most stunning disruptions in the established model of economic development. Although it was a crisis in the business cycle, essentially a crisis in the rate of growth, the lingering effects of the crisis, particularly in the financial sector, caused great tensions in the economy.[12] These tensions impinged enormously on public finances, as much at the provincial level as at the national level. As Marcelo Carmagnani has pointed out, "the 1890 economic crisis would allow the Argentine ruling class to complete a financial plan elaborated in the 1880s," which was most concerned with "reducing the dependence of public finances on the commercial connection existing between the Argentine economy and the world economy, with the precise objective of making exports more competitive" and "achieving a slow increase in earnings coming from activities not directly linked to foreign trade."[13] Such a plan favored a notable increase in indirect taxation in the form of taxes on consumption.

The taxation question would become one of the socially most explosive issues in the two decades following the 1890 crisis, just as it had been in Buenos Aires when, at the end of the 1870s, the first attempts at such a transformation in public finances had been attempted.[14] Contemporary with the impetus given to indirect taxation, in Santa Fe there were new attempts at direct state taxation on commodities that begin to show profits: grains, the flour milling industry, and *quebracho.* The producers of each would make their own contribution to a social protest, which would slowly acquire political significance, creating at the same time a complex articulation between citizenship and taxation, between social demands and political demands that seek expression through certain forms of association, or even in revolutionary practices.

Voting, Taxation, and Municipal Space

While in the spheres of political decision making at the national and provincial level there was being orchestrated a fiscal policy that would incorporate all citizens (in accordance with their capacity for consumption) for purposes of the state's upkeep, at the local level these same "contributors" saw undermined their decision-making power over fiscal resources by virtue of being deprived, in their status as foreigners, of the municipal vote in the majority of towns in the central-west and southern parts of the province. In others, municipal government was itself disappearing, giving way to the *comisión de fomento,* or the "Development Commission," virtually subordinate to the decisions of the provincial executive. It is for that reason that the year 1891 begins with three key problems looming over both the areas of agricultural colonization and the southern port city, Rosario: the recovery of the municipal vote, naturalization, and taxes on grains.[15] These concerns are going to weave a complicated connection between the role of the taxpayer and the citizen for the *colono* and also cause great tensions in Rosario's urban setting, affected by the exclusion that its propertied citizens found themselves subject to, a class greatly augmented by an immigration that had stimulated the city's growth.

The municipal vote for foreigners, elective justices of the peace, and the rejection of cereal taxes became the battle standards in the agricultural colonies. Mobilizations and journalistic harangues constituted the first modes of social protest.[16] The Esperanza newspaper, *La Unión,* for its part, assumed programmatically its role as shaper of public opinion:

> Our manifesto is directed to the *colono* who, abandoning the plough for the moment in order to remind himself he is a citizen, expresses in simple language his dislike of arbitrary behavior and disregard of his citizenship; it speaks to the hardworking, upright foreigner who finds himself dispossessed by the arbitrary rule of special treatment and exemptions protected by the law. ...[17]

The social actors quickly established organizations between 1891 and 1893 that went beyond strictly social concerns and had political repercussions. Both the *Centro Político Extranjero* (*CPE*) and the *Unión Cívica* (*UC*)—two organizations of national importance—emerged as spokesmen for the demands of this predominantly foreign-born population.[18] As with the later *Unión Cívica Radical,* both the *CPE* and *UC* were organized as the local affiliates of two national organizations. The actions of both reminded the political class of the necessity of listening to the concerns of an emerging sector of Santa Fe society and reaching some solutions. In the case of the *UC,* the objective would become to incorporate such concerns within a broader political program. In the case of the CPE, it would be to convert the concerns into a source of mobilization for the groups it represented.

Contemporary with those organizations there appeared, especially by 1893, certain kinds of sectoral "protection societies" and regional "units" ("unidades") that, confronted by a welter of demands, directed their efforts mainly at dealing with the taxation issue, an issue that was charged with political meaning given the political circumstances of the moment. Thus, both in their meetings and in their armed rebellions (to be discussed later), the taxation problem pits these organizations against their traditional "enemies": the justices of the peace,[19] against the excesses and abuse of power in which taxes "mesmerize and vanish," but also against their own consular representatives:

> The Swiss of the Progreso colony have directed a statement of protest and a demand for redress to the Swiss minister in Buenos Aires, requesting the removal of consular agent, Enrique Queller, the government commissioned tax agent.[20]

At the same time, the debate over taxation led to a consideration of two central issues: how to spend the obtained revenue and how to oversee spending. As *La Razón* noted in February 1893, imprisoned *colonos* in San Gerónimo asserted that they "were generally disposed to pay the tax as long as the money was used to pay schoolteachers, elective justices of the peace, and local public officials in general; they asked also for a restoration of elective municipal government."[21] These demands clearly presented a vision of what the *colonos* expected from taxation.[22] By their way of thinking, the revenue ought to be directed toward the necessities of the community, for the common good. In this regard, education and the legal system were priorities. At the same time, they defined the space in which the fiscal policy that affected them should be settled: the municipality.[23] Taxpayer and citizen became one and indivisible; the categories were fused through the demands expressed and the actions undertaken. Even in the organization that the immigrants formed in the southern part of the province, the "Society of Taxpayers," an organization driven to a great degree by groups from Rosario,[24] they claimed among other objectives, "to study the right that foreigners have to vote in municipal elections."[25] Analyzing the problem, the organization quickly submitted to the legislative bodies the demand for a constitutional reform to "give the foreigner the vote."

In this period when the social actors from one or another region alternated, as resistance tactics, pacific protest, and "revolution" (rebellion), certain forms of consciousness were developing that put specific problems at the front and center of the debate, problems that had heretofore either been ignored or considered of secondary importance. The center stage, emerging in a complex process, was occupied by local problems, although without losing sight of the strong articulations that the local maintained with provincial or national concerns. Among those problems that had been relegated to a secondary plane by the foreigners who lived in the agricultural colonies or in Rosario was, without question, the issue of naturalization. In reality, the national debates that began in 1887 on this issue had had little resonance among them; but the deep tensions that emerged in 1890 revived its importance.[26] It is obvious that—faced with obstacles to political participation and confronted with the decisions of public authorities that impinged directly on their economic interests—the search for the status of citizens through naturalization responded to the necessity for a greater insertion in an organic institutional structure within the norms of the prevailing political system.[27] In this way, naturalization became one more element in the struggle for the municipal vote and in opposition to the grain legislation. But it was an issue that opened new questions and established new tensions.

The issue of naturalization brought into play not only questions of identity but also practical concerns that resulted from being citizens of a foreign nation, such as unquestionably were concerns about the protection afforded foreign consulates or the exemption from certain "patriotic duties" in their adopted country that were viewed in highly negative terms (military service, for example). It is for that reason that in their political meetings in both the countryside and city, foreigners looked for an alternative to naturalization: the acquisition of political rights without abandoning their nationality. The proposal was taken up as much by immigrant associations[28] as by the *CPE* and the *UCR*.[29] These latter two, anxious to win the support of the immigrant community, of becoming to some extent their spokesmen and of mediating between them and the provincial and national authorities, joined forces in a common platform. Although it remained a merely programmatic statement that could not be translated into reality, it would continue to strengthen important bonds of solidarity among their respective leaderships.[30]

An issue of national importance such as naturalization achieved a singular resonance in Santa Fe once it was linked to the struggle for municipal space, a municipal space within which it was sought to define, among other things, how revenue acquired through tax collection would ultimately be spent. The fight for municipal control between the provincial executive and the *colonos* or members of Rosario's urban bourgeoisie led to a certain redefinition of the municipality that, in the case of the *colonos,* becomes clear around 1894. In the entire preceding phase, it was not only from above that the municipality had been given an administrative definition; such a definition was also shared by the social actors themselves in the municipal setting. As late as November 1891, the foreign delegates from the *colonias* declared:

> We ask for the municipal vote. First, because community power is not a political issue, as some mistakenly believe, but one of self-government, in essence popular and democratic, concerned with the welfare, health, safety, and family organization of every people. Second, because we are taxpayers and residents of the municipality. Third, because, since we take an active part and have duties in collective life, we should also have "rights." Fourth, because foreigners have been the founders of the agricultural colonies and for that very reason we desire to administer our own interests, sweat, and labor, all the more so since self-government had been promised to the founders of the *colonias* in special contracts, signed by the national authorities, with full power to do so, contracts which still are in our hands.[31]

This is a declaration with a strong "administrative" content, consistent with the standards of order and governance established by *roquismo* and with its core the "possible republic" that sought to exclude from the decision-making spheres those citizens outside the elite, outside the better bred. But within that declaration a certain tension is revealed. These taxpaying residents, who have duties in their collective life, are demanding their "rights." These are the demands that will henceforth be widening the breach between the political elite and the social actors, and will later permit a spokesperson from one of the agricultural colonies, *La Unión,* to refer to municipal government in the following terms:

> The primary political entity... where rights and civic duties are born... (according to)... the great contemporary thinkers, is municipal power, a natural outcome of the federalist principle... which nurtures republican sentiment.[32]

This issue of municipal government that had dominated three long decades in the life of the Santa Fe municipalities undoubtedly assumed greater urgency as the result of the real life experiences of the rural and urban residents in social terms and their subsequent effects on the political context. The assemblies, the mobilizations, the permanent resort to the right to petition, are stages in a protest that will finally culminate in the 1890–93 revolutionary cycle. But the perception did not turn out to be as clear for all the social actors as it would be for the *colonos* from Esperanza and Rafaela. Rosario's social and political dynamic especially was very different from that of the agricultural colonies, and the differences are clearly related to the characteristics of those fractions of the bourgeoisie who lived in the city and on whom a strong impression had been made by the liberal opposition groups opposed to the status quo. Among them, for many years factionalism was favored over democratic practices. Nevertheless, the shared, frustrated experience of the *Partido Constitucional* and their involvement in the Radicals' armed uprisings were opening rifts from within that would slowly grow wider.

I should note, however, that despite the differences that divided *colonos* and Rosario's bourgeoisie, they will both attempt to consolidate their positions in the local political space, seeking to generate there new alliances at the regional and national level. If for some more than others "politics and democracy" clearly loomed as pending questions in order to achieve the "true republic," for everyone a stage was beginning when one issue would become critical importance in the struggle to reformulate politics: the organizational issue.

The Experience with Building a Party: the *Unión Cívica* and the *Unión Cívica Radical*

The changes experienced in the way of understanding and conducting politics, culminating in the crisis of 1890, unquestionably had ramifications in the party system. In the previous decade, just one party, the *Partido Autonomista Nacional* (*PAN*), the same one that served to keep the country's provincial elites in close contact, established its political hegemony. The 1890 rebellion called into question both the hegemonic role played by the PAN and the philosophy underlying its existence: indisputable and undisputed unanimity. Just a few years later, some of the principal leaders of the *roquista* order would recognize that one of the most significant outcomes of the 1890 crisis was the emergence of a strong and organized opposition. Echoing these sentiments, in the debate over the party platform at the *PAN*'s 1897 convention, one party member stated:

> We renounce the impossible dream of establishing a government that satisfies everyone. That was, Mr. President, the concern of another era. This urge for unanimity is what led to catastrophe and we, Mr. President, repudiate unanimity, we know that opposition is necessary, that not only should it be tolerated, but that it is essential to allow for its participation, to use it, as the British do, as an indispensable element of government.[33]

The opposition that was recognized in 1897 as necessary was the one that in the upheaval of revolutionary struggle in 1890 ushered in a new kind of political organization, one removed from the sinecures of the state and based on grassroots committees in Buenos Aires province as conceived by Hipólito Yrigoyen. The Buenos Aires *Unión Cívica,* seeking to become a national organization, adopted its definitive structure with individual party affiliation as its starting point, establishing a formal party apparatus that would seek, as did the former *Partido Constitucional* in Santa Fe during the 1883–84 conjuncture, the establishment of a grassroots consensus. In the 1891 party convention held in Rosario, party ideologue Leandro Além defined the concept of the party as a "civic" one by proposing:

> The *Unión Cívica* was not, is not, and cannot be a party such as those that formerly dominated our political movements; by its very nature and by the mission it has established, it must reject and combat all personal agendas, all the self-interested ideas of cliques, all secretive and unseemly schemes, all deals that are calculated to make a mockery of the people's vote, thereby cheating the great hopes that Argentines nourished after all the efforts they undertook to recover their rights and rescue a humiliated public morality. The *Unión Cívica* wants and looks for ... a pure, genuine, free and independent people's opinion in the Republic, whose autonomy is one of the first inscriptions on its flag, at odds with centralizing tendencies whose effects we have just suffered ... it calls for the cooperation of all the provinces by means of their free and legitimate deliberation in order to choose the candidates for higher office in the nation, ones who deserve their trust and embody their aspirations in these solemn moments.[34]

The party charter established, step by step, a political structure that through neighborhood or district committees, once the secret vote was established, slowly took shape in a process of selection and representation. These grassroots organizations and more routine party assemblies sought to preserve the autonomy of each sector in the party. According to one observer, these organizational guidelines sought to the give the *UC* "... a permanent organization, guided by principles and essentially impersonal."[35]

For the first time, a party proposed not only to express the popular will, but also to guarantee freedom of conscience by protecting the voter from the pressures of a public vote, of the voice vote, a common practice in the preceding political tradition. At the same time, and from another perspective, the proposal reclaimed for the agenda of the 1890s, to some degree, the old project dating back to the origins of the nation, of the *Partido de la Libertad* and its principles of a national liberalism, the party that *mitrismo* (present during this conjuncture as well) had attempted to consolidate. Although at the time of the fall of the Rosas dictatorship in 1852 the organizational structure of the party was not given priority, the concept of the party "as a collective community" as a depository of political loyalties rather than an instrument of the state or of the leader did take priority, as it did over the pretensions of the political leader "to represent all of society" and "to express all legitimate political aspiration."[36] In light of this objective—to be the "cause" confronting the "regime"—it became difficult in 1890, as it had been in 1852, to formalize precise programmatic positions. Nevertheless, some meaningful aspirations that were being expressed throughout these years in society were realized. In a gathering to commemorate the events of July, 1890, it is Francisco Barroetaveña (one of the principal leaders of the *UC*) himself who, facing those very actors who came from a long tradition of struggles, echoed their concerns:

> The program of the *Unión Cívica* offers to all the provinces, municipal government, the right to vote, provincial autonomy, just as the constitution prescribes, not as Roca and Juárez Celman have cynically practiced it... it offers, in a word, administrative integrity and severe punishment against electoral fraud and the misappropriation of public money.[37]

One additional element was added to a program that at the same time drew distinctions between it and Mitre's project. Although the new party sought to break with elements of the past and to "invent" its own tradition that distinguished it from the rest, in its objectives, strategies, and daily practices, it used the past as a source of legitimacy and to build its own party structure.[38] Using the fragments recovered from its history in Santa Fe, we are able to analyze some facets of that "invention."

The "invention" to which I refer is that which offers as a symbolic founding moment the 1890 uprising, even though it is undeniable that the uprising also had close ties with the recent past, ties not only with previous party projects (the *Partido de la Libertad's* in the national arena or regional expressions such as the Buenos Aires *Partido Republicano* in the 1870s or Santa Fe's *Partido Constitucional* in the 1880s), but above all with those ways of "doing politics" that, excluded or marginalized in formal politics, had been growing stronger over the course of several decades. A basic element of the new organization was that of placing value—from a democratizing perspective—on the natural space in which politics "are done," the public plaza, the "forum":

> The phenomenon which one notices has its legitimizing origins in the patriotic reaction of 1890, and is the exclusive work of the *UC*

> party. ... Before, political meetings had a clear tendency: when those essentially democratic acts were organized in some Argentine "forum," they were broken up by the gunshots of the thugs in the service of those in power and the cohorts of *autonomismo*. ... After that memorable event (the 1890 uprising) the people had a vision of the future and came to the "forum" whenever the leadership of *civismo* called them. ... Even the *PAN* now looks favorably upon the practice of public meetings in which the leaders of a political movement address the people regarding political platforms and present and future problems.[39]

The selection of the public plaza, "the place of harangues," as a political space that leads the way to political participation, abandoning the "the darkness of authoritarianism," entailed establishing another strategy for this new political style. The latter not only was characterized by a participation that sought to be broad-based but also by the introduction of certain cultural practices. In that regard, together with the political meeting, there are employed the "civic processions" that these "fellow disciples," "these fellow citizens" inaugurate as ways of keeping alive the memory of the deeds of their leaders. The "civic processions" that the newspapers recount, particularly between 1890 and 1895, combined old and new traditions. Such processions were old in the sense that these actors, generally excluded from formal politics by virtue of their foreign status, traditionally had resorted to mobilizations in order to make their demands known to those in power, marching through the streets, literally voicing their concerns through public protest. These practices were new, however, in the kinds of demands being made in public demonstrations, above all in their intent to establish a party tradition.

This novel character is revealed by the interest expressed year after year in recovering events and leaders whose memory must be preserved. In this way, July and September become key months to which February was eventually added (all months in which there took place Radical rebellions), in the political calendar. Além and Aristóbulo del Valle were leading figures in the party hagiography who were revered and rendered homage after their deaths; and the dates of their deaths became important milestones in party life. Also novel were the characteristics of the marches (political clubs parading with their own banners, bands of music, the supervision of public order, etc.) and the presence of people clearly identified either by social class, generation, gender, or ethnic origin. The presence of women constitutes a revealing and novel example. Those who had been eternally absent in the decision-making process of public policy regarding the family, though ever-present in the articulation of political alliances, appear in the public arena (as they had in previous social protests, especially in the *colonias*), as much as in the very days of rebellion (when they took care of their husbands, collected money for those in prison or blacklisted) as in the celebrations (cheering the protesters on, taking up collections to decorate the plaza for demonstrations or prepare the protest banners, holding banquets in honor of the leaders, etc.).[40]

Although the grassroots protagonists of the processions are the political club—a traditional organization—their participation has a novel character. Though it is true that among the Radical clubs in Rosario there can be detected some factions that make reference to traditional public figures (the clubs Mariano Moreno, General San Martín, General Paz), the majority of the clubs present their own leaders (Captain Eloy Brignardello, Leandro Além, Colonel Arnold), some of whom die in the Radical uprisings, or events in which these leaders participated, as the symbol of their group.[41] This concern for the "construction of a collective memory" was not reduced to sporadic events such as marches and demonstrations, but also had in the party press, a press that becomes more widespread in Santa Fe, a flexible instrument to both reconstruct and publicize the trajectories of their leaders (especially the newspaper, *La Bomba*) and to fashion public opinion through their editorials.[42] So significant is their role as shaper of public opinion and so deep the commitment to their party that throughout the decade, but particularly between 1893 and 1894, that newspapers are closed for undertaking "agitating propaganda" and for "their incendiary and riotous propaganda."[43] In the case of others, the number of copies allowed to be printed is limited, publishers are jailed, or circulation is obstructed.[44] The newspaper, *La Bomba,* reports:

> The Editor-in-Chief and other correspondents who tour the countryside are personally entrusted with attending to the complaints of our subscribers against the postal service and its employees who, failing their duty, obstruct the circulation of *La Bomba* in one way or another.... It is for this reason that, paying heed to the reports we have received, in addition to the abuses that he have seen with our own eyes, the Rosario district has responded to our petition and issued summons against the heads of the postal services of Gálvez, Ballesteros, Arroyo Seco, and Pueblo Ramallo.[45]

Society did not remain passive in the face of such pressures and mobilized. In both the *colonias* and Rosario, public opinion expressed its opposition to restrictions applied to the press, and the social actors either petitioned or marched in protest, displaying their willingness to defend free speech.[46] This public opinion in the process of consolidating itself had a clear sense of the right to express its demands in the public arena; it is for that reason that, beyond immediate or parochial concerns, diverse social actors came out in the defense of such civil liberties, transcending even party loyalties.

Social Actors and Politics

In April, 1890, with the establishment of the Radical political club, *Juventud Cívica Rosarina,* the *Unión Cívica* initiates a slow process of party-building in the province. Among both the governing party and *Unión Cívica* itself, there existed a consensus that the recruiting grounds of the latter were to be found fundamentally among the foreign *colonos* and the Rosario bourgeoisie. Thus, in the correspondence that Gálvez sends to Roca in August, 1890, it is asserted:

> With regard to the *Unión Cívica,* it is a plant that will not sink deep roots in this province and even though in Rosario it may be able to count on some foreign elements, we will snuff it out with the creole and with the law, under the protection of a liberal and prestigious government.[47]

The distinction made between creoles, or native born, and foreigners appears constantly as part and parcel of the clientilistic system that the traditional ruling cliques erected in the public sphere. For that reason, *La Razón,* a newspaper that supported the UCR, emphasized from Rosario that the true Radical opposition was to be found there since the city of Santa Fe (capital of the province of the same name) was comprised of government (hence native Argentine) employees. At the same time that it described how state resources were used to establish political clients, it minimized the potential support for reform from the *colonias* since there "people worry more about working than getting involved in political disputes."[48]

It is undeniable that the newspaper's interpretation was highly partisan, indeed possibly linked to a broader plan to assert the hegemony of Rosario within the *UCR* party structure. Contrary to such interpretations, there were clear links between the party and the *colonos;* namely, the explicit defense assumed by the *Unión Cívica* and then the *UCR* in the tax revolt, the struggle for elected municipal government, indeed the very right to establish such a government.[49] Such a defense was not only expressed in writings and petitions, but in marches and public demonstrations in which the demand for these things was constant. In the *UCR* party "Manifesto" addressed to the people of Santa Fe and published July 30, 1893, certain attitudes are revealed related specifically to the cause of the *colonias:*

> Revolution is not a simple cure applied uniformly to the evils resulting from the deficient and irregular exercise of the institutions that a sovereign people may have given itself; but when everything has been corrupted and it is useless to attempt to recover through the law what has been lost, revolution remains the last and drastic resort which it is necessary for the citizenry to take and those people who are not resigned to live under the eternal yoke of illegality and crime.... The suppression of the free vote has converted the executive and legislative powers into nothing less than *de facto* usurpations. The judicial powers dependent on both, in the same way. Government functionaries are chosen without considering their qualifications to perform effectively in the positions they were awarded; monies are misspent and squandered without any control; the *Banco de la Provincia*, established at great sacrifice, has been the scene of the greatest immoralities in the handling of its money and today is hopelessly bankrupt; the public debt has risen to enormous levels, (the province) has defaulted on its payments, and taxes have increased incessantly to the point they have become intolerable, provoking resistance in the case of certain ones, such as the grain tax, which have justifiably caught the attention of the entire country.[50]

This deep questioning of the provincial political and administrative system created a space for the kinds of demands dear to the world of the *colonos* but also to Rosario's bourgeoisie. The participants in the agricultural colonization process, for their part, continued to affirm their commitment to the new party. After having been leading actors in the events of July and September, 1893, harshly repressed and persecuted after the failure of both revolutions (particularly in Esperanza and Rafaela), the *colonos* played an important role in the formation of Radical party committees.[51] At odds with the image of clannishness that the Rosario press sought to portray, and deeply involved in social issues and Santa Fe politics, the *colonos* began to analyze the problem of naturalization from the point of view of participation. From the convocation of Conrado Hang in 1891 when they took over en masse a meeting of the CPE "... with the harvest tasks completed, the cause of citizenship for purposes of contributing their greatest and best efforts to the country's reorganization ... " they had made much progress.[52] Although one cannot speak of a "massive" outpouring, it can be noted that, despite pressures from the foreign consuls and the preaching of some of the ethnic press, there was an increase in the demands for naturalization in the province's central-west zone, especially between 1894 and 1895.[53] The phenomenon must have assumed some degree of importance to the extent that a pro-government newspaper, *El Orden,* included them as part of the Esperanza opposition and states, "It is a fact that the Esperanza Radicals will participate in the election next Sunday tied to a democratic front made up of naturalized foreigners."[54]

If these are the actors in the *colonias,* naturalized and nonnaturalized foreigners, what is the makeup of Rosario's Radicalism? The "civic procession" carried out July 30, 1894 described by *La Razón* allows us to not only assess the numerical weight of each group but to also detect the nature of Rosario's Radicalism. Thus it is affirmed, for example, that the *Club Gral. Teodoro García* possessed an "executive committee comprised of more than 100 people from high society" or that the club *Capitán Eloy Brignardello* is made up of "420 elite Radicals" or finally that the *Club Mariano Moreno* has among its 350 members "the cream of Rosario's youth."[55] It is undeniable that members from the worlds of trade, finance, and the liberal professions, analyzed by me in another work, played an active role in the *UCR.*[56] The great majority shared with the agrarian sectors their status as foreigners but unlike them do not seem to have been concerned about naturalization.[57] Some of them, moreover, were militants of the former *Partido de la Libertad* and allied with a faction opposed to *autonomismo,* the local incarnation of the ruling party, the *PAN.*

In 1891, when there occurred the rupture between Mitre's forces in the *Unión Cívica Nacional* (*UCN*) and the supporters of Além organized in the *UCR,* it was only the unfolding of the first act in a conflict that continued to gather force for the next seven years, and in which two other figures emerged: Bernardo Irigoyen and Hipólito Yrigoyen. Both men would bitterly contest control of the *UCR* at a later date. In the dynamic of Santa Fe politics, such leadership disputes and the party splits gave rise both to a spatial displacement of the actors and to a rebirth of longstanding tensions that had caused bitter disputes in the past. Through their respective histories, both political parties seemed to have staked out their bailiwicks in the province:

> The *mitrista* party which presently goes around disguised as the *Unión Cívica Nacional,* has few supporters in Rosario but many in the provincial capital ... the Radical Party, on the contrary, has many adherents here in Rosario but few in the capital. A government drawn from this party would not be able to function there.[58]

The Rosario newspaper, embarked on a campaign to move the provincial capital to Rosario, forgets once again the standing that the *UCR* had achieved in certain *colonias.* What becomes clear is that, despite overlooking this influence, dominant in its characterization of the *UCR* is the close relationship that the party maintained with liberalism. The liberal platforms opened the road to a model of development in which these actors felt themselves to be the "mature fruit," the same platforms that explained their differences on the Church and education. These platforms served as a basis on which to forge an alliance between groups that had been gaining strength over the course of four decades: the provincial capital's liberal faction, the *colonos,* and fractions of Rosario's bourgeoisie. It is for that reason that when *El Rosario* tried to highlight the differences between *alemistas* and non-*alemistas,* it did so by highlighting the liberal-clerical opposition:

> There is in the Radical party two antagonistic tendencies that only can camouflage their differences and avoid manifesting their reciprocal repulsion, through constant struggle against a common enemy.... The youth, in the majority liberal, is just as it is everywhere, the generation which rebels; and with their leader Além, it seeks to move forward along the path of modern civilization, leaving its imprint on the government a spirit of absolute freedom and true regeneration. On the other side, the clerics disguised as *cívicos* and protected by doctor Irigoyen, will never accept the revolution's conquest of ideas, seeking always to undermine institutional progress.... The old guard, that is to say the clerical wing, seeks a government of priests under whose pernicious influence previous administrations have been corrupted; in the meantime, the healthy segment, that one comprising the liberal nucleus of a vigorous youth, begins to demonstrate their desire to move the provincial capital to Rosario because the *rosarinos* are lovers of freedom.[59]

After the failed attempts in the 1870s to make Rosario the nation's capital, the new actors would continue demanding a political recognition for the economic and social importance the city had achieved. The southern part of the province, with Rosario in the lead, sought in this new stage to play a hegemonic role through a party emerging out of the crisis. The rifts opening in the heart of the *Unión Cívica Radical* grew deeper as a result of the dispute over abstention or participation. The conflict did not take place therefore between the provincial capital and Rosario, but between the latter and the area of the *colonias* in the central-west, especially Esperanza, and in the very heart of the port city itself.

Abstention versus Participation (1890–1896)

Throughout the three decades preceding the events of 1890, Esperanza's *colonos,* whether naturalized or not, had given clear signs of their intent to control municipal government. Whether displaying a democratic logic or involving themselves in factional disputes, they had throughout these years a history of active participation that led them, on the one hand, to reformulate their idea of what constituted politics, and, on the other, to view outside alliances as an excellent way to project their demands into the public domain. The process was long and difficult, but they did manage to articulate their roles as producers, tax payers, and citizens. Possibly for that very reason, and because they systematically prevented the governing party, despite its attempts, to gain a foothold in the *colonia,* is why they objected to the idea of abstention.

Rosario's situation was different. From the days of the organization of the *Partido Constitucional,* the liberal opposition had enormous difficulties in competing electorally.[60] The actions of the political bosses and of the justices of the peace from within the well oiled party machinery of the *Partido Autonomista* were entangling this opposition in its own contradictions and establishing limits to its actions. The bitter struggle in which the Constitutionalists saw themselves involved, their helplessness in the face of the maneuvers of *roquismo,* their incapacity to devise an alternative strategy to the clientilism and factionalism of the *PAN,* led them to abstentionist practices during the decade of the 1880s, practices that were adopted again in the early 1890s by both the *UC* and *UCR.*

As in other periods, representative voices of Rosario society, coming now from the executive council of the *Unión Cívica,* showed their determination not to participate in elections and sketched out a familiar scenario:

> Today we decided not to participate in them (the elections) in order not to lend legitimacy with our presence to the illegal procedures that have been used and are being used.... The forced confinement of citizens of this city and of the countryside, the strategic stationing of military troops and armed groups comprised of unsavory individuals for purposes of carrying out premeditated crimes; obstacles placed in the way of the *Unión Cívica* party in order to make impossible its participation; the brazen falsification of the electoral roll; the depositing of hundreds of ballots in the urns before the voting even began, reaching such a point of having 800 votes by the opening of the polls; the supreme obligation of our party to save lives that should not be uselessly sacrificed; and finally the belief that the province has lost all possibilities for honest elections.[61]

The position adopted in the summer of 1892 turned out to be a recurring practice of the Rosario *UC,* and then *UCR,* in those years. The belief in the absence of an alternative to the governing party at election time appears to have been a shared sentiment only in the *colonias* in that February of 1892. At that moment, it is recorded that the "partido cívico," that which held in Esperanza and almost the entire department an "absolute majority," boycotted the election "in order not to sanction with its presence the brazen fraud carried out by the governing party. The various elements comprising the *colonias* show a similar position.... All together they have only been able to get to the polls some fifty miserable farmers, a number of whom have been brought by justices of the peace or political officials from the district."[62]

The *colonos,* however, rapidly regained municipal control in the *colonia* and encouraged and then supported in 1894 the candidacies of Carlos Bosch for provincial deputy and Eduardo Yost for senator in their electoral district.[63] Such a recovery occurred immediately following the two revolutionary crises of 1893 and then the violent repression that followed them. It is undeniable that the failure of those movements, the deportation and jailing of leaders, caused as much at the national as at the provincial level a turning inward by the party. Nevertheless, the *colonos* showed in that conjuncture an effective capacity to react—the result possibly of more consolidated organizational practices—had permitted them to get their representatives into the provincial legislature. From that point on, the voices of Yost and Bosch are constantly raised in defense of the municipal vote and attacking the grain tax.[64] Although they did not achieve their objectives, they did influence some changes in the ways the tax was collected, attempting to temper the taxes' inequities and the deleterious effects on their constituencies.

The different evolution of the political practices in the *colonias* compared to Rosario, despite a press favorable to the UCR that awarded the city a privileged place in the party, can be looked at from another angle. An important part of the Radical leadership in Rosario had emerged from sectors of the elite who, over the course of many years had alternated control of the municipal government. These people had not been absent either from the factional politics and clientilism or the corruption of forms of popular representation practiced by the *PAN.* They stood as a leadership qualitatively different from that of the *colonias* and with a varying degree of commitment to their constituencies. It is this dimension of the problem, the very one that the month previous to the first uprising in 1893 reappeared in the press reports, which revived the longstanding issue of corrupt politics. Through this press coverage it is possible to perceive the difficulty that existed in this urban setting for "constructing" the citizen. So difficult does it appear for the citizens "to demonstrate in some fashion that they live and think," that the newspaper sought to discard this mediating space that the party constituted in favor of a citizenry identified with their roles as taxpayers:

> The law-abiding citizen must take advantage of the precise moment of weakness of the political parties for electoral undertakings and rise up and occupy the abandoned terrain, showing with his actions his intention that politics and government henceforth be professional and honorable.... Because into the general disaster that has fallen all our institutions has also fallen the majority of our men of leadership, criticized nowadays for their lack of credibility.... Everything is just a question of taking the initiative. Of course, it is necessary to prevent political groups from taking it. It belongs to the leaders of taxpayers' organizations and the leading social organizations alone to make respectable the call and offer local venues to sign agreements. Each social organization will have to lend its prestige to one or a number of candidates, the ones who would be accepted or rejected in the caucuses until a final slate is filled.[65]

El Rosario assumed the role as spokesman for a segment of civil society that, despite the existence of a party with certain modern characteristics, seemed not to have felt represented by such a party. The ethical content underlying suspicion of the parties and of the leadership is clearly represented in the platform that is proposed for the so-called "Taxpayers' Union." This organization strongly criticized the link between politics and government patronage in the pubic bureaucracy at the same time that the state increasingly assumed the role as the administrator of community organizations. The problem of public spending—its control and decisions on how it was spent—was proposed as the basis to rally the representatives of the "community" who, at the same time, would permanently seek to influence the decision-making process:

> They will link up with popular organizations, seeking to join forces for purposes of charity, public health, education, and all that is capable of the common good, striving always towards the greatest possible degree of perfection.[66]

As in other municipalities, the represented citizenry established a set of priorities in the handling of the public space for the common good. Yet tellingly, the "Taxpayers Union" that had "appeared not affiliated to any party" turned out to be permeated with Radical groups, as one member affirmed:

> It seems that the slate (in the "Taxpayers' Union") which has come out triumphant in the vote is identical to the one that before had been agreed upon by the executive committee of the Rosario Radical Party.[67]

This led the newspaper to make an appeal to the Radical leadership that it "withdraw its candidacy and vote for the taxpayers' slate," opining that if "there ends up being only 100 Radical voters in Rosario, the influence that this party aspires to have later on ... is going to be very weakened."[68] Beyond the position assumed by *El Rosario* in defense of the taxpayers, the nature of the appeal gives room for thought. In the first place, it is clear that there existed in the city a leadership that was not disposed to surrender its authority, an authority that it traditionally controlled despite certain occasional setbacks, even though it never managed to transcend its local character and have resonance at the provincial level. By the same token, it is reasonable to think that the party is, for these actors, only one of a number of possible mediators between state and society, being able to deploy other means of mediation. In those other movements or local organizations, the status of citizen is inextricably linked to that of the taxpayer.

In this experience of June–July 1893, neither the Radicals nor the members of the "Taxpayers' Union" participated in the elections. That led *El Rosario* to parody the situation, on the verge of another revolutionary uprising, in the following terms:

> This house is for rent.
> Without voters or candidates.
> The year is 1893.
> The house is unoccupied
> because of a lack of candidates for municipal office
> and because of a lack of voters.[69]

Opposition opened, then as in 1890, the road to revolution. The 1893 revolution took place in two months, July and September. The July revolution had its epicenter in the city of Buenos Aires and was completely suppressed. The September uprising, though it sought like the July rebellion to attain national dimensions, was centered in Santa Fe. It too was crushed. Nevertheless, the Santa Fe uprising was more successful by virtue of the ability of the revolutionary group there to control the provincial government for some 21 days. In keeping with the broad demands of the revolution, Santa Fe's civic-military rebellion once again was organized around the motto of *sufragio libre,* seeking to put an end to the electoral fraud and the general political corruption that characterized the provincial governments' handling of the public patrimony. There rallied behind the *Unión Cívica Radical,* not only the *liberal* factions of the local ruling class but also important sectors of Rosario's bourgeoisie and some of the *colonias,* especially Esperanza and Rafaela. The ultimate failure of the two uprisings brings with it, as previously noted, repression and disarray among the participants. Nevertheless, the party committees and delegations sought to keep alive their traditions and practices without entering the electoral struggle. The participation that, under other party denominations, characterized the *colonias,* is manifested in an abstentionist position in Rosario between 1895 and 1897. In these years, such mainstays as Víctor Pessan, Rafael Ferrer, José Castagnino, Luis Pinasco, and Agustín Landó disappeared from the various municipal councils.[70] Nevertheless, this period is one characterized by an active process of party reorganization and internal debate.

The year 1895 began with political assemblies, particularly in the departments of "Las Colonias," "Capital," and in "Rosario," for purposes of electing delegates for a party convention to be held in the Federal Capital in March of that year. In accordance with party statutes, the province would be represented by four delegates.[71] Slowly the Radical committees formally selected their representatives; but internal tensions caused new confrontations. The key problem that affected the Radical party structure resided once again, at both the national and provincial level, in disputes over the strategy chosen to confront the ruling class. Two roads were presented in the debates: that of abstention and that of a confederation of opposition parties. While at the national level the dispute became personified in three figures—Leandro Além, Hipólito Yrigoyen, and Bernardo Irigoyen—in a struggle that took place particularly between the national party committee and the party committee of the province of Buenos Aires, in Santa Fe the *colonos* were pitted against groups from the provincial capital and Rosario.

The *colonos* did not cease in demanding throughout these years the granting of municipal status for their communities and the right of foreigners to vote, going so far as presenting to the public authorities the argument of the benefits to the provincial treasury that would result from taxes collected in those areas where the number of municipal residents, and therefore of taxpayers, increased.[72] Consequently, they were not disposed to submit to a policy of abstention. It is for that reason that, at the beginning of 1896, the delegates representing the "Las Colonias" district are expelled from the Radical Party executive committee after refusing to accept the abstention call. The arguments posed by the *colonos* were now rooted in the idea that participatory political practices nourished democracy while self-exclusion did not.

The opposition to abstention became more complex following the emergence of a national campaign to form a common front of opposition parties. In Santa Fe, the proposal, agreed to by members of the leadership in the provincial capital and in Rosario, meant taking up again the alliance with *mitrismo,* a prospect that the leaders of the *colonias* opposed:

> The cliques are finally satisfied, those which call themselves opposition parties, since the central committees of Santa Fe and Rosario have found to be patriotic, the proposed "confederation of parties," accepting its overall purpose and what is more, conceding representation to the "Unión Cívica Nacional" which is almost a nullity since we doubt it would scrape together 200 supporters around its banners in our province, leaving themselves covered with ignominy for such a shameful "agreement."[73]

For the *colonos,* the alliance served no purpose, since "the people are exclusively Radical and the pro-government elements that could counteract their collective force in legal elections are insignificant."[74] The debate once again reveals as much the distance between this leadership and the province's two most important urban centers as it does the uneven insertion of Radicalism in each area. This is clearly expressed in the editorials that the newspaper *La Capital* publishes on January 22 and 23 of that year, editorials that are echoed in *La Unión.* These editorials not only sought to recover a hegemonic role for Rosario "where the sentiment of public opinion beats more strongly," but they also made suggestions about how to formalize "a single party center." In that regard, the newspaper appealed at first for a recognition of the leadership from the capital city grouped together in the *UCN,* made up of "men of fortune and property" and "with sufficient influence to guide the popular will towards honest, uplifting, and patriotic goals." At the same time, it warned the Radicals that "alone and isolated, they will never move beyond their present limits unless they ally, differentiating themselves somewhat in their way of thinking, from those who pursue the same goal and defend the same cause—to put an end to public nepotism and government by family in this unfortunate province."[75]

The newspaper justified such an agreement by the potential contribution each leadership could make. Just as the *UCR* showed a great capacity for social mobilization, so the *UCN* could offer the means whereby this mobilization achieved its objectives. The problem presented as primordial in its editorial was one that had scarcely appeared on the scene before:

> A thousand or ten thousand citizens on public streets means nothing... without those to get the funds to pay for meetings, trips, tickets, the support of party activists, for printing party propaganda.... And since the founding fathers of Radicalism were the last representatives of a moneyed liberalism, and the party stalwarts are working men who cannot give themselves the

luxury of unbridled spending, what other road remains open than unification and alliance with those who, though few in numbers, have available what others lack?[76]

Radicalism faced, as with any modern party, the need to acquire resources essential to maintain and make viable its organizational structure. However, the bottom line was that one or another party leadership did not show many differences. Contact was ongoing, before the formation of the *UCR* and following the rupture. The same newspaper, outspoken defender of the Constitutionalists and the *UC* in the 1880s, and then the *UCR* in the 1890s, is clearly moving in the direction of becoming a spokesman of the *UCN*.

The *UCR* leadership did not see real obstacles for achieving unity, did not "feel" itself to be very different. Rather, it was fractions of the leadership, and even more so the rank and file, who ended up highlighting the differences. It was the presidents of the party clubs, those who are leading the opposition, who questioned the February 22, 1896 formation of the opposition committee. The pressure exerted against the Radical committee in Rosario led to the resignation of the president, Joaquín Lejarza.[77] Lejarza himself recognized in his writings the "signs of decomposition" beginning to be felt in the *UCR*. He attributed "the divisions and the discord" to a disregard for the hierarchies at the various levels of the party leadership. He noted the preeminence of "the authority of the presidents of the local party clubs over that of the central committee" the latter of which could not be considered the "patrimony of *caudillos*." The question was, to what extent did the highest bodies of the party leadership—the central committee and the assembly—appear controlled by sectors of a leadership that did not respond to rank and file demands?

The crisis unleashed in the heart of the *UCR*, which had led at the national level to the resignation of Além as party president, not only had particular repercussions in Rosario but also in the *colonias*. There, in Esperanza and Rafaela, followed by others, both abstention and the coalition of opposition parties were resisted. But only Esperanza seemed disposed to participate in an electoral struggle to confront the government party. Under the banners of the *Comité Democrático* (the Radical name could not be used), the Esperanza Radicals faced off against the *autonomista* and Catholic coalition. The latter managed, through fraud, to impose its own candidates. Nevertheless, these Radicals remained prepared to confront their adversaries in the public space and through party competition. The nickname of "intransigentes" or "rojos" with which they were tagged would seem to have put them close to the faction headed by Hipólito Yrigoyen. Nevertheless, the *colonos* opposed Yrigoyen's abstention tactics.

The Party and its Practices in the 1897–1898 Conjuncture

Both the progovernment and the opposition press in Santa Fe dedicated some of their news columns to cover the debate that, in the heart of the parties, revolved around objectives and platforms. What is noteworthy is that this debate occurred during a stage of upheaval in party structures; a period of reorganization and reformulation but also of the emergence of new ways of doing things.[78] The *alemista* ideology, taken up by the provincial newspaper, *El Municipio,* defended as basic party objectives the formation "of the citizen, the consciousness of his rights and feelings of solidarity in a common destiny," an altruistic position that seemed to have moved even further away from a party reality in which splits and tensions had multiplied.[79] Thus, at the national level, Roca's candidacy in the 1898 presidential elections came as a shock for both the Radicals and the *UCN*. Behind the reiteration of demands for governments upheld by "public opinion," as expressed in "free" elections and guaranteed by agreed-upon electoral reforms, *alemistas* and *mitristas* would recover discursively the problem of the party.

La Capital already following the line of the *UCN,* published Mitre's reflections in the meeting of the *Plaza de la Libertad:*

> Effective politics must be impersonal, understanding for impersonal that which takes its inspiration above all from the collective sentiment of the public good, subordinating to that purpose all interests, because this is the only thing that justifies the existence of organic parties of principle in a republic.[80]

This perception that so many had of the party, in actual practice was weakened in the deals struck by the leadership and the priority given to personal and factional interests above "the common good." Such a situation allowed Carlos Pellegrini, a former president and a leading *PAN* figure, to make political use of the situation, attempting to present party challenges to the established political order as a contest between the common good (the nation) and private interests (Buenos Aires):

> The true formula then is the following. The nation gets nothing, and the capital rules (because it has the nation's monopoly of intelligence and culture).[81]

At the same time, promoting a reassessment of the PAN as "the first truly national party," established by Avallaneda via the unification of provincial "localisms" and the "localism" of Buenos Aires, Pellegrini sought to minimize the influence of the opposition. While the *UCN* was presented as the "heir of the party of freedom," but which had not managed to penetrate the masses, the UCR—whose party base came out of the former *autonomismo*—was presented as "more an attitude than a political party."[82] In Santa Fe, however, where *situacionismo* was wracked by internal struggles, afraid of not being able to guarantee a transfer of power, split between *leivistas, galvistas, iriondistas,* and an opposition that never managed to define a clear position on abstention or participation, the emergence of the candidacy of Juan Bernardo Iturraspe worked as a true binding force that narrowed the distance between the government party and the opposition.[83] Within a matter of days, the Santa Fe *UCR* witnessed the national party convention's decision to participate in the elections, despite the bitter opposition of the party committee of Buenos Aires province and the subsequent loss of one of the party's most outstanding leaders—and opponent of the Buenos Aires committee—Santa Fe's Lisandro de la Torre. De la Torre opposed the Buenos Aires committee's pretensions to hegemony within the *UCR* but supported the policy of a coalition of opposition parties.[84]

Nevertheless, an intense energizing of party life occurred, as much for purposes of analyzing the issue of the coalition of "popular parties" as for its internal institutional reorganization. In October 1897, Santa Fe's Radical convention resolved to eliminate the central party committees in the provincial capital and in Rosario and to give greater powers to the departmental committees. At the same time, it assumed a position favorable to the candidacy of Iturraspe. Although it did not pledge its electoral support—something that becomes clear days later after voting in favor of abstention—it did declare a possible collaboration if "this one (Iturraspe) realizes in the government the UCR's political and administrative reform."[85] The position of the convention regarding electoral abstention clearly was not shared by some sectors of the party in the provincial capital and in Rosario, but above all in the *colonias.* It is noteworthy that the candidacy of Iturraspe caused an intense mobilization in the province, expressed as much in the press—that public space and transmitter of public sentiment—as through the welter of petitions and demonstrations of support for the candidate that appeared throughout the province.[86]

The case of the *colonias* displays its peculiar character once again. They consistently opposed the coalition policy, but it is the figure of Iturraspe who triggered the reaction. On the one hand, this was due to the fact that Iturraspe was someone closely tied to the process of agricultural colonization and, through blood relations, to important families in that colonization.[87]

On the other hand, it was partly due to simple political loyalty, the fact that some of these *colonias* had given their political support to him previously. In the decade of the 1880s, Esperanza, forming the "Unión Provincial," had played an almost solitary role in the Iturraspe's first electoral race. Though Iturraspe lost the election, the large turnout demonstrated the degree of consensus reached by the candidate. Finally, the support was due to the fact that Iturraspe was acquainted with, and at particular moments shared in, the struggle for municipal government. It is precisely there we find one of the keys to the support that he would receive from the *colonos*. His triumph would return to them after some ten years of struggle, following a new reform, the right to vote.[88]

Clear signs of the degree of acceptance of the candidacy are, on the one hand, the rejection of abstention and, on the other, the search for a way out without destroying the party. Acceptance is also demonstrated through electoral participation as independents.[89] The pressure exerted on the party apparatus must have been sufficiently intense and deep for *El Tribuno* of Santa Fe, a Radical party organ, to communicate to the provincial party convention that:

> Upon responding to the telegrams of petition sent to them, thus were the Radicals authorized to work in behalf of that candidacy and lend their support, in the manner they deem appropriate, as long as they did not abandon the party and respect the decisions of its representatives... that they could not oppose a candidacy that, in itself, was a guarantee of good administration and a firm promise to establish free elections, respect for the law and for institutions.[90]

The party defended its banners and contributed to an alliance sui generis that brought together representatives from the government party and the opposition around a figure who emerged as a guarantee for a future political accord. The triumph of Iturraspe demonstrated with clarity that it was really not the party apparatuses but rather the actions of different sectors of society, some of which were independent of the party machinery, who were the architects of Iturraspe's electoral victory. Provincial Radicalism could not avoid its own tensions and those existing at the national level. Neither was *situacionismo* easily able to operate. The final years of the decade left important questions pending: the question of increasing struggle regarding political representation between the provincial capital and Rosario and the recurring problem of the requests for mediation between civil society and the state.

Disintegration, Reorganization and Revolution. The Difficult Road of the UCR 1898–1905

The rifts caused by the coalition policies, as well as by the persistence of abstention calls posed in the interior of the *UCR,* produced, from 1898 onward, a strong disarticulation at both the provincial and national level. The wear and tear resulting from the tensions existing in various parts of the province, between leaders and the party rank and file, led to an atomization, to an organizational unraveling. Some leaders went their own way, as in the cases of Mariano Candioti, Lisandro de la Torre, and Aldao y Carlos Gómez. Others reasserted their positions on electoral abstention, despite having achieved for their foreign constituencies the right to the municipal vote.[91]

Only in 1904 can one speak of the beginnings of a reorganization of the Radicals. Their clubs resume their affiliation drives, and the national leadership's ongoing contacts were reestablished with party committees in the process of reorganization. The different positions, however, remain in effect. Thus *La Capital* related, for example, the presence of the most intransigent current in the party's interior, the so-called "Red Radicalism":

> The latest tour of the distinguished citizen leaders of Radicalism known as the "reds," because they do not enter into agreements, pacts, alliances with other factions, has been beneficial for that group as there has been reestablished in the provinces the committees which were disbanded and they are proceeding rapidly to organize a rank and file which increases daily due to the return of old and loyal members. ... The revolutionary idea is giving way to the idea of party competition.[92]

The newspaper was excessively optimistic with regard to the electoral participation of the UCR since the latter continued to insist that the electoral option lacked guarantees, despite the electoral reforms that are realized from 1902 onward. Nevertheless, it is true that the UCR continued its political work, thereby ensuring the functioning of the entire party apparatus. The revitalization of party life produced in some places such as Rosario a genuine internal party democracy, introducing on the national scene certain rank and file members such as Ricardo Núñez or Ricardo Caballero.[93] Although the party did not abandon its position on abstention, once again in 1904 its rank and file and some of its leadership became involved in the March congressional elections.

This involvement was enough to entice the participation of a former party member, Lisandro de la Torre, who offered his candidacy for the

9th district in the March, 1904 elections. During the campaign, de la Torre's famous "letter-platform" touched again upon the problem of tax reductions and cutting excessive spending.[94] Facing de la Torre, who was backed by Radical clubs, important Radical leaders of Rosario, and "popular groups," were two candidates: Pelayo Ledesma (who competes with de la Torre for the seat in the 9th district)—supported by the *Sociedad Gremial de Troperos y Propietarios Unidos*—and Luis Lamas, endorsed by businessmen and men of social prominence in Rosario. Ledesma and Lamas also received the support of the important Radical clubs General Roca and Manuel Quintana.[95] The *situacionismo* faction of the party tied to the *PAN*'s national project reappeared in this instance. The various factions of Rosario's bourgeoisie divided their support and were represented by several spokesmen. The former strongman of Radicalism, Lisandro de la Torre, received the explicit though not exclusive support of sectors of the party who adhered "in their character as Radicals" to the elections, despite the fact that the party abstained. The other two candidates were backed by social organizations and the typical clubs of the former *situacionista* factions.

Fraud was once again the instrument through which the government party imposed its candidates. With the support of 24 percent of the registered voters in the southern district (961 votes), Luis Lamas was elected to congress, while Pelayo Ledesma achieved the same results with 34 percent of the votes in the northern district (938 votes against de la Torre's 593).[96] The fraudulent practices of the past obviously had been only superficially changed with the promulgation of the new electoral law, which required voter house registration and the maintenance of a permanent electoral roll. The voter house registration (*empadronamiento domiciliario*) was supposed to put an end to fraud by making voter registration the responsibility, no longer of the local courts, but of specialized government civil servants who compiled the voter lists by personally visiting every home and registering qualified voters. In actual practice, fraud simply moved from the courts to the registration bureaucracy. Such a situation supported the positions of the abstentionists, who observed with growing unease the revival of old practices to guarantee electoral control. The press seconded these feelings in its criticisms of the provincial governor:

> And so he only seems worried about incorporating the party *departamento* leadership into the electoral machines, in giving them responsibility for organizing fraudulent or stolen votes, or votes obtained through threats to persecute law-abiding rural citizens, people who do not live off of politics nor off the public payroll, but from their work or savings.[97]

The dichotomy between politicians and citizens, between the behavior of the representatives and the represented, appeared once again. But there also reappeared two perennial problems that had existed since 1890: that of coalition tactics and that of revolution. The policy of coalition promoted by some political and social sectors brings with it a novelty in the 1905 electoral campaign: the voice of a new social actor, the trade union movement. This actor, whose press had chronicled a long history of struggles, begins to express its opinions in nonunion newspapers. It was once again *La Capital* that through editorials signed by one Pancho Guernica, identifying himself as a worker, introduced the working class into politics via the press. What is significant about Guernica's efforts was not so much his vindication of a coalition project but above all his reflections on the changing positions of "the working and industrial classes" with regards politics:

> Until now the working and laboring classes, encouraged by the indifference and condescension of the most exalted classes in terms of their culture and social position, cowered by the forces of the government party who used the police as a substitute for debate, and above all, evasive and reluctant to become involved in politics, have persistently stayed away from the ballot box since they did not recognize any legitimate interest represented in election results. The workers did not believe in elections. They have never known that the power to elect governors, congressmen, and mayors is theirs too, those who toil away in the country's principal workshops and industries, the ones that are its lifeblood. Today, everyone realizes the importance of the right to elect their leaders and it is becoming understood the duty which is incumbent upon all citizens in this respect.... They are already realizing that what they called politics, reviled as wicked and in which they felt like outsiders, believing it to be an area reserved to those favored by fortune, something that was barred from the statutes of all the working class organizations as a pernicious germ threatening its health, they now understand that politics is a sacred obligation of every citizen, an unavoidable, inalienable right that elevates to the same level as much the men who carry loads on their backs as those who live off government sinecures.[98]

Although this new perception among the working class was not as great as the journalist suggested, his words did convey a situation in which the political space was widening and being observed from different perspectives. Contemporary with this process, the revolutionary strategy reached an ebb tide; and as with its predecessors, the uprising of February 5, 1905 was a new frustration with its outcome of repression and party disarray. As in the rebellions of 1890 and 1893, the protagonists of the revolution of February 1905 were an alliance of civilians and the military united behind by the leadership of the *Unión Cívica Radical.* As in those "revolutions," there were similar objectives and the rebellion experienced similar obstacles and outcomes. In Santa Fe, the revolutionary groups suffered the same fate as their counterparts in Buenos Aires and Córdoba: they were swiftly crushed, despite the success of the Cordobans in capturing President Figueroa Alcorta (with whom they eventually negotiated their surrender). The failure of the 1905 rebellion meant the continuance of old problems without resolution, new and old actors searching for tactics and the appropriate means to employ in the public space. Perhaps in this process, 1908 looms as a relevant conjuncture to see how the social actors and the politicians, with all their tensions and contradictions, behaved in Santa Fe.

1908: The "Liga del Sur" between Social Movement and Party

From the moment of its emergence, the *UC,* subsequently the *UCR,* had sought to become a political alternative. To that end, it adopted different strategies that, in turn, caused rifts, breakaway factions, and even some instances of membership insubordination overwhelming party structures. In the final revolutionary endeavor, the party revealed its limitations to become a genuine shaper of the "public will" in the heart of a society in which social movements and the manifestation of the popular will in a specific act—voting—were now attempting to head in the same direction. The "possible republic," on the other hand, sought clearly to separate the spheres of political power and the experiences of social life. It sought to exclude important social actors from participation in the making of decisions. It blocked, or at least limited through a restricted representation, the appearance of a series of social demands in the public space.

For its part, certain social actors also, as I have noted, had attempted to distinguish between daily concerns and politics, the latter conceived not only as a space foreign to them but also as something corrupt and tawdry. If at one stage they gave priority to social action over politics, the decade of the 1890s led them to reformulate this perspective without still modifying their attitude toward electoral participation. The years following 1890 showed that the party that presented itself as an opposition, the *UCR,* had not managed to rally around it the different groups through which such an opposition was expressed. From society, voices were raised and demands were made that the party did not have the ability to accommodate. Its bases had widened, a generational reshuffling had occurred among the leadership, but groups of those citizen-taxpayers and consumers who formed the base of the party did not feel represented. The *UCR*'s appeal was then directed to the social organizations that formed, in turn, party organizations at the local or community level in whose leadership notably stood out Radical leaders. The question then was, did the problem reside in the abstentionist line of the party and the impossibility, by not participating in elections, of placing the demands of the group in the public space? Or rather, these demands that basically refer to problems of a fiscal nature that affected them as taxpayers or consumers, did they reflect collective necessities that were beyond the abilities of the party leadership? Or finally, were the demands to be advanced through other channels of mediation of a nonparty nature in the search for a more "effective" relationship with the state?

Perhaps going over the experience of 1908 may help find some answers. For Santa Fe, as for Córdoba or Entre Ríos, 1908 looms as a moment of great tension and a high degree of social and political mobilization.[99] Unlike previous decades, the central area of Santa Fe's agricultural colonization does not seem to have been strongly affected by the unrest. On the other hand, the southern part of the province, with Rosario and Casilda in the lead, played an especially prominent role. The root of the problem lay as much in the province of Santa Fe's fiscal policies as in those of the municipalities. During the 1890s, the three categories which made up some 95 percent of provincial tax collection were the following: direct contributions such as sales taxes (which oscillate between 25 percent and 28 percent of the total), licenses (*patentes*) of various sorts (between 21 percent and 23 percent), and the "impuesto del papel sellado" (between 45 percent and 46 percent), a tax paid for any kind of notarial or bureaucratic procedure between private citizens and the state and that included the mailing of farmers' almanacs of various kinds. Within this last rubric was also found the much reviled grain tax (the tax on grain sales notarized by public authorities), which already in 1894 represented 23 percent of the revenue collected under this tax. The fiscal apparatus restructured in the country in the 1890s, geared now to the internal market and domestic taxes, experienced successive adjustments that sought to bring up-to-date property values and tax assessments by taking into account the complexity and diversity of commercial and industrial activities.[100]

Between 1904 and 1908 the taxpayers' mood was altered not only by the introduction of two new taxes on productive activities (the taxes on *quebracho* and its derivative, tannin, as well as on flour milling), but also by tax increases, despite public opposition to them.[101] Although the two new taxes did not comprise a significant percentage in the total amount of tax revenue collected, they became a new bone of contention among groups who in 1908 had to contend with the closing of numerous establishments, such as in the case of the millers[102] and the expulsion of the labor force from the *quebracho* plantations, faced with yet another of their industry's devastating business downturns.[103]

To the provincial demands were added those of the municipalities, where license fees had become the principal source of fiscal revenue. In early April, the first signs of tension were observed. In Santa Fe's provincial capital, the public plaza again became the setting where opposition to municipal policies regarding taxation would be expressed. The petitioners, backed by more than 1,000 "respectable signatures," asked for the repeal of the municipal tax code. Although they felt themselves called upon to protest in their status as "taxpayers," their criticisms extended to the whole of municipal government. Led by the city's merchants, the mobilization apparently had its main support in the retail sector, whose voice—according to its declarations—"is that of the capital city consumer"[104]; a voice that, if ignored, could convoke movements similar to those that had occurred in other provinces such as Córdoba, where full-scale rioting took place. Something similar, although without producing mass demonstrations, took place in Rosario. There, the *Centro de Almaceneros,* "expressing that displeasure felt among the union of retail members" protested, demanding a lowering of the taxes on licenses that had been doubled, noting at the same time that without a favorable response, the situation "was tending towards the ... eruption of a latent public protest."[105]

The circumstances of that moment show the social actors and politicians working along parallel paths, paths that through the actions of some members of the leadership establish certain points of convergence. While Radicalism reaffirmed its process of reorganization with public demonstrations or in the internal reform of the party apparatus, sectors of the society that the *UCR* was appealing to were searching for answers to daily concerns. At this stage, despite party reorganization, they appealed for the construction of a "community party." What was the objective of this community party?

> A strong community party is one by virtue of is numbers and the caliber of its supporters in which politics, properly speaking, has sensibly been put aside, and has no other concern than that of purging a sickly organism. ... Such a party's first acts accord with a general aspiration, perfectly reasonable, of an equitable reduction of taxes that had been increased with arbitrary criteria lacking also goals of the public good. ... There is, then, in the city of Santa Fe, a public spirit, there is an independent opinion which is capable of forming worthy, sovereign bodies in defense of principles and common interests.[106]

The proposal encapsulates two fundamental premises. First, discursively it seeks to distance the community party from "politics," but desires to occupy that other very contested space of political decisions that are its "administrative" and "policy-formation" functions, in order to settle a crucial issue of power: tax policy. It elects to do so, after employing instruments of social protest such as demonstrations, petitions, the independent press, and through a political organization: the party. In the second place, it calls attention to the problems of "general interest" or "common good," highlighting once again that apparent dichotomy that seemed to take precedence within the *UCR* between the specific realm in which politics are manifested—elections—and the kind of politics expressed in people's daily concerns. There is also present the problem of abstention, a situation that would hinder the possibility of action and of realizing established objectives. This would justify the participation in elections by leaders committed to the movement and even by the party membership.

In later documents, sentiments in favor of recapturing a political participation of a more traditional kind are revealed in positions taken on the issue of the system of local representation. Those "sovereign bodies" consider that, "One does not find in the executive council the community's true representatives. . . . The remedy is familiar; since the citizens do not have the right to elect who administers what is "theirs"; let them elect, at least, those who control their actions from the council."[107] The demand to exercise, to regain their rights, is clearly expressed.

The emergence of a community party is repeated as an experience in diverse Santa Fe municipalities—as in the case of Casilda with its "Unión Popular"—as well as in Entre Ríos and Córdoba.[108] Such a party is considered as a direct expression of the "general aspirations," removed from the "corrupting world of the old style politics in use," and with the exclusive objective "of keeping a watch on the correct handling of public affairs."[109] But once again this is not the only organizational form that the popular sectors adopt, since in some places it is substituted by *juntas* or defense leagues, or by similar associations. Nevertheless, the idea of the party as mediator had not been lost. The problem resided, in part, in the fact that this idea was not associated, by some sectors, with the *UCR.* Possibly, as *La Capital* expressed it, certain sectors of society sought a greater articulation between party organization and daily concerns, a desire that had been voiced over the course of almost two decades:

> After the public demonstrations emerge the parties which are latent but shapeless at the time of the events. The parties have genuinely serious

> and viable platforms when they are based on the economic concerns of collective life.[110]

The general programmatic declarations of the old *Partido Constitucional* or of the *Unión Cívica Radical* were insufficient for these taxpaying citizens. In this light, the reflections of the delegates at the 1897 *PAN* party convention appear visionary:

> Should we raise some economic issue in our banner? ... Economic problems are still looming on the national horizon. We are beginning to feel the weight of their importance on the legislative agenda; we predict that in a short time it will be impossible to govern this country without openly addressing these questions; but, in the meantime, the situation is only just beginning.[111]

The actors who lead these movements, in their majority, come from fractions of the small and middling urban and rural bourgeoisie, people with clear aspirations of social mobility and concrete demands, demands related to conditions for economic growth and capital accumulation. The economic demands of 1908, unlike what had occurred at other moments, were not to be found encapsulated in liberalism or protectionism, but rather in the dual proposition of taxation and the return of revenue to society.

Faced with a party that had its sights set basically on the institutional and electoral dynamic, the social movements paved the way for a new organizational structure, conceived as the synthesis of all previous undertakings. In November 1908, those actors who had sought to assume a representation for society, which basically is associated with the "taxpaying and non-taxpaying people," establish "the new *Liga del Sur* defense party." The manifesto that gave justification for the party's establishment noted that "it is not a question of creating an essentially political party, but of a league or group of national and foreign citizens who, being all direct or indirect taxpayers, long for the progress of the land where they are undertaking their activities."[112]

These citizens limited their efforts to the "interests of the center and south of the province" even when they believed that "no one should be excluded." They asserted that "the cause is great and therefore encompasses all those in the province's *colonias,* towns and cities who have been signaling the enormous injustices which governments and representatives appear to enjoy perpetrating against all those bountiful regions which work and produce."[113] While the Radical Party central committee threatened to expel from its ranks those members who joined the new party,[114] the *Liga*'s call won the support of "the respectable conservative classes," of the "independent press," of "representative elements from trade, business, and labor."[115]

Farmers, landowners, merchants from Rosario and numerous provincial towns—the core of southern Santa Fe society—showed their support. Rapidly, despite its reservations about party forms, the *Liga del Sur* acquired a party structure and the establishment of party committees and departmental subcommittees was expanded. Certain leaders, such as Daniel Infante, attempted to deepen along these lines the "genuinely democratic" representations, with aspirations that the committees become "truly the delegates of the membership." Although the organization was still incipient and fragile, the party attained a clear importance in the southern part of the province. Why the south? The visible head of that south was Rosario, the nucleus in which liberal fractions were supported after the 1868 revolt, the one in which the platform of the *Partido Constitucional* of the 1880s was sketched out, the program that nourished the *UC* and then the *UCR*, attempting to achieve a hegemonic role which would permit it to displace the government party. De la Torre's statement to the Radical Frugoni Zabala leaves no doubt:

> The South was already prepared for this great reaction. The North was not. The facts demonstrate it. Successes like that of the *Liga del Sur* are not by chance, do not happen unless public sentiment trembles with enthusiasm for the idea.[116]

What separated the leaders of the *Liga del Sur*—among whom stand out former Radicals such as de la Torre, Lejarza, Landó, Pessan—from their previous party affiliations? For Frugoni Zabala, the Radicals cannot "but look upon with sympathy almost the entire program of the *Liga*... that program which agrees almost completely with the one that *in concreto* the Radical Party committee that has its headquarters in this capital has promised and propagated in the northern and central departments of the province and which like the other satisfies the deepest and most steadfast aspirations of both the rural and urban populations in the province. It perhaps only differs in the means by which it bring them into the practical realm."[117]

In a fashion, the Radical leader hit the nail on the head. In reality, the *Liga*'s program resembled the proposals to defend the municipal autonomy which Radicalism had been advocating since its origins in Santa Fe, though there was a basic distinction. Beyond its influence in the southern part of the province, the *Liga* emerged as a result of an alliance between local groups, an alliance to be controlled for and by the latter. Radicalism, subordinated to a national party structure, had to operate within certain parameters and with a leadership who frequently did not take into account local interests and concerns. The *Liga* appeared as the result of a series of social movements led by a new leadership that was not disposed to surrender what power they held, aspiring to expand it even further while retaining control.

The south, especially Rosario, reaffirmed at this moment its decision to end the subordination it had been subjected to over the course of decades in terms of political representation. The final act that had affected the central dynamic of the provincial political economy (whose economic locus had moved since the middle of the 1880s to the south) had to do with the departmental reorganization of the north and the refusal to modify the situation in the south.[118] In order to put an end to its marginalization in decisions of power, these fractions of the commercial, financial, and agrarian bourgeoisie in the south proposed a series of substantial modifications in Santa Fe's institutional framework.

One of the primary objectives was to establish political legitimacy on a broader base, introducing representation for minority parties and conceding the vote to foreigners. Although the latter proposal was an attempt to respond to one of the most heartfelt demands of the immigrant sectors, it did not represent a total opening up of the political system but a limited one.[119] The second demand—proportional representation—was intended to break with the government's monolithic structure. The instruments chosen here were the municipal governments, proliferating throughout the province. Unlike the artificial departmental structures that privileged some areas over others, the autonomous municipalities—at the same time that they favored government decentralization—permitted a greater local government intervention in all aspects of civil society: education, the economy, the justice system, the control of the forces of public security.[120] The new political dynamic also could not leave untouched the problem of the provincial capital, all the less so since the *Liga* had plans to undertake an economic development policy that sought to broaden the possibilities for participation by groups heretofore scarcely represented. It is for that reason that it ended up proposing moving the capital to Rosario. The problem of the capital, with all its social and political implications, became, in turn, a breaking point for the power arrangements worked out between groups in Rosario and in the provincial capital city. Neither of the two leading sectors was prepared to surrender. Not only are the groups tied to the government party in the provincial capital opposed to the proposal, so are the Radicals there, as asserted by Frugoni Zavala who considers that "to modify the constitutional provision that established the capital in Santa Fe" was "an unacceptable idea any way you looked at it."[121]

If the rupture of the *UC* and the emergence of the *UCR* and the *UCN* had implied a particular shape for the provincial political map, as much from the point of view of the political space as from that of the political actors, the emergence of the *Liga del Sur* precipitated significant readjustments with respect now to the UCR. Not only can one see an exodus to the new movement by traditional Rosario leaders (Castagnino, Pessan, Lejarza, Landó, Ricardone, Ortiz, and Araya), but also by some outstanding figures from the local committees of the *UCR,* among whom would have to be added members of social organizations that until then had had a marginal participation, and youth groups who were just beginning political activity. Even the world of agricultural colonization that allied with the *Liga,* revealed some new actors, certain groups of southern *colonos,* headed by those from Casilde, not a few of them small sharecroppers. Although this sector shortly tempered its support, during the first stage it was deeply involved with the *Liga.*

The *UCR* seemed to retain strongly its constituency in Santa Fe's capital city and the *colonias* of the center-west. The leadership from the provincial capital, beyond its involvement in the community party that never managed to supplant the government party, reaffirmed its adherence to Radicalism. Among the *colonias,* Esperanza continued to be the great Radical redoubt, the one where, despite the pressures, electoral support was assured.[122]

The year 1909 put to the test the degree of consensus achieved by the *Liga* as much at the level of social mobilization as in electoral participation. On January 3, the *Liga* organized a demonstration to protest an increase in the provincial budget for that year and the tax measures that accompanied it. In the manifesto directed "to commerce and the people," the *liguistas* asserted that "The increase in taxes for 1909 fall exclusively on the working man. The burdens are not proportional to one's capital nor to the income of the contributor but fall rather on one's work. The southern region will pay for the increase because it is the most hard-working, the most enterprising, and the most dynamic."[123]

The fact is that neither the *liguista* leadership nor the region were disposed to accept the fiscal imposition, all the less so if to this was added the new municipal taxes recently decreed. For this reason, a strategy of classic social protest was employed: the general strike. Beginning January 9, Rosario's businesses closed their doors, the stock exchange itself shut down, joining the protest. The tension in the streets, calls for the mayor and members of the city councils resignations, led the municipality to temporarily suspend the new provincial tax ordinance.

The unrest did not abate and on February 6, while the *UCR* undertook demonstrations commemorating the 1905 uprising, the *Liga* supported the strike call proposed by a general assembly of the bakers', shopkeepers', butchers', and other unions as long as the mayor's office refused to abolish the high municipal taxes. As social tension increased, the city was paralyzed. The movement widened with the support of other sectors of the working class.[124] The revocation of the ordinance nonetheless came too late. The municipal council resigned en masse and the new council president appointed by the outgoing council members and entrusted with addressing the protesters' demands was a prominent member of one of the fractions of the Rosario bourgeoisie and a *Liga* follower: Santiago Pinasco. Although social tensions cooled off due to some temporary solutions to the situation, the problem would nonetheless persist. The opposition movement to the municipal and provincial taxes erupted that same month in other urban centers in the province. In the June elections, the *Liga* would capitalize on the consensus achieved on the taxation issue. With a party ticket that included the names of its principal leaders and those from organizations such as the *Centro Unión Almaceneros* and the *Centro Unión Dependientes,* the *Liga* managed to prevail in the municipal elections, a political space that it would tightly control for many years.

Final Remarks

The process that unfolded over the course of almost two decades and that has served as the focal point of my analysis allows for some final reflections. In the first place, there undoubtedly appeared on the scene in these years the actors who would be direct participants in the social mobilizations and also in the political organizations that emerged out of such mobilizations. We are dealing here with the world of a small and middling bourgeoisie who were attracting other sectors of society, summoning them to place their demands in the public space. These actors, however, displayed complex and at times contradictory behavior. While those who are located in the original settlement region of rural colonization in the province's central zone, with its heart in Esperanza, refuse to accept the pretensions of the *roquista* state that they become autonomous with regard to the rest of society and are prepared to run (in agreement with Sarmiento's proposal) the "double risk of politics and democracy," the same does not happen with Rosario's property-owning classes.

Those from Esperanza, who over the course of several decades had been developing in the local setting the idea of the municipality as a political body, undertook in the 1890s actions in pursuit of the acquisition of full citizenship. In that sense, they undoubtedly worked, both in the experience of adopting Argentine citizenship and in choosing a party, to mediate between civil society and the state. It is this context in which, following the brief role played by the *Centro Político de Extranjeros* (*CPE*), the *UCR* emerged as a valid response, with important degrees of consensus.

Rosario provides a place for incorporating new perspectives. Its "propertied classes," those who also over the course of time had been voicing their demands in the public space, resort to different forms of mediation, oscillating between social movements and party structures. The *Liga del Sur* itself, unlike the Radical Party, showed in its organizational stages that ambivalence, an ambivalence that cannot be reduced only to the political organ that is the party, and which emerges basically from the conception that those sectors have of the citizen and of the citizen's most immediate sphere of action—to the extent that the others turn out to be precluded by the dominant sectors—the municipality.

The citizen to whom this sentiment would eventually be appealing, was not then convinced of its "legitimacy of origin," did not yet have in the 1890s a clear idea that the municipality in which he resided drew together all the powers of the state. But it is undeniable that the idea was going to go on charting a course that led it to establish the basis for a new legitimacy. One of the circumstances that favored this process resides in fiscal policy. Alberdi's "possible republic" established the dichotomy between "political rights" and "civil liberties." Nevertheless, the exclusion of civil society affected these "civil liberties" guaranteed by Alberdi's prescriptions for democracy and echoed by the Generation of 1880. The contradiction was revealed in one of the areas where the decisions of power operated meaningfully: fiscal policy. It is for that reason that fractions of Rosario's small and middling bourgeoisie, involved in the movements and organizations that defended free suffrage, also began to formalize their demands to participate in policy decisions that sought tax reform. To the degree these social actors made advances in that area, moving from the demand to administer revenue to a political agenda, the tension between the realm of civil liberties and that of politics grew. That taxpaying citizen who formed the bedrock of the movements of 1908–09 was arguing for things subsumed in that tension, and in the search for new responses to escape it.

Translation by James P. Brennan

Acknowledgment

The author would like to thank Tulio Halperín Donghi and Hilda Sábato for the comments and suggestions offered for this chapter.

Notes

1. In 1890, the first signs of discontent erupted. The economic crisis of that year unleashed a wave of popular protest whose repercussions would linger for the remainder of the decade. In May 1890, members of the *Unión Cívica,* in alliance with disgruntled sectors of the military, organized a rebellion against the central government. After the suppression of the revolt, the *Unión Cívica* was reconstituted as the *Unión Cívica Radical* (*UCR*). Armed rebellions were attempted again by the *UCR* in 1893 and 1895, both ending in failure. In the case of all three uprisings, the province of Santa Fe was an epicenter of the unrest and protest.
2. A provincial perspective on these years provides precisely the kind of opportunity to recast the national story that this volume suggests is in order. Viewed from Santa Fe, these years certainly appear more ones of social effervescence, political experimentation, and meaningful change than has often been thought. A classic book of a foreign scholar can serve as a point of reference: David's Rock study of the *Unión Cívica Radical, Politics in Argentina, 1890–1930: The Rise and Fall of Radicalism* (Cambridge: Cambridge University Press, 1975). Rock stresses the continuities of the most important political reform movement of these years with the liberal oligarchy's national project and downplays its novel character. Though there was unleashed, according to Rock, a fierce struggle for the spoils of political office, the fundamentals of the system were never seriously questioned. I maintain that not only were they questioned in Santa Fe, but that this questioning had a transforming effect on social relations and power arrangements at both the local and national level.
3. This chapter represents a continuation of research I have been engaged in for a number of years. See Marta Bonaudo, Silvia Cragnolino, Elida Sonzogni, "Discusión en torno a la participación política de los colonos santafesinos. Esperanza y San Carlos (1856–1883)," in *Estudios Migratorios Latinoamericanos* 9 (1988); "La cuestión de la identidad política de los colonos santafesinos, 1880–1898. Estudio de algunas experiencias," in *Anuario* 4 (1988); and Marta Bonaudo and Elida Sonzogni, "Redes parentelas y facciones en la política santafesina, 1850–1900," in *Siglo XIX. Revista de Historia* 11 (México, 1992).
4. This same undertaking is to be found in a number of recent works that seek to reconsider the issue of citizenship and political participation beyond electoral politics, trying to reassess those hypotheses that stress the political apathy of important sectors of the population, especially the immigrants. Among other studies should be mentioned Hilda Sábato and Ema Cobotti, "Hacer política en

Buenos Aires: los italianos en la escena pública porteña, 1860–1880," *Boletín del Instituto de Historia Argentina y Americana* 2 (1990); Hilda Sábato and Elías Palti, "¿Quién votaba en Buenos Aires? Práctica y teoría del sufragio, 1850–1880," in *Desarrollo Económico* 119; Ema Cobotti, "Mutualismo y política. Un estudio de caso. La Sociedad Unione e Benevolenza en Buenos Aires entre 1858 y 1865," in *L'Italia nella societa argentina,* ed. Devoto Resoli (Rome: 1988); and Hilda Sábato, "Ciudadanía, participación política y formación de una esfera pública en Buenos Aires, 1850–1880," *Siglo XIX. Revista de Historia* 11 (Mexico, 1992) and especially her recent, *La política en las calles: Entre el voto y la movilización (Buenos Aires, 1862–1880)* (Buenos Aires: Sudamericana, 1998).

5. Hilda Sábato, "Ciudadanía, participación política y formación de una esfera pública en Buenos Aires, 1850–1880," p. 70.
6. Tulio Halperín Donghi, "1880: un nuevo clima de ideas" in *El espejo de la historia* (Buenos Aires: Sudamericana, 1987), p. 248.
7. Notalio Botana, *El orden conservador* (Buenos Aires: Hyspamérica, 1986), p. 66. "Roquismo" is the term used to denominate the political order that begins in 1880 and culminates in the consolidation of state power during the presidencies of Julio Roca (1880–1886, 1898–1904).
8. Bonaudo, Cragnolino, Sonzgoni, "La cuestión de la identidad política de los colonos santafesinos: 1880–1898," p. 263. The municipality was an urban administrative organization with responsibility for public health, the police, and other activities directly related to daily urban concerns. In the beginning, the municipalities also had responsibility for the courts and education, though they now lost control of those powers. This was the political space where those who were not part of the provincial power structure could most directly exercise political influence. The municipalities were established for communities that had a determined number of inhabitants, going from a minimum of 1,500 inhabitants in 1872 to 4,000 or 5,000 in 1883. Settlements that did not meet this minimum number of inhabitants could not have a municipal government and instead were administered by a *comisión de fomento*, which had authority over administrative affairs but whose members were appointed by the provincial executive. The *departamento* was a larger political-administrative organization that included a number of cities and smaller communities and generally was directly dependent on one figure in the central provincial government: the provincial executive. The divisions by *departamentos* allowed the provincial executives to maintain a centralized power structure. Moreover, the representation by provincial congressmen and senators was apportioned by *departamentos*. There were two senators per *departamento* while the number of congressmen (*"diputados"*) depended on the population of the *departamento*. These *departamentos* were gerrymandered to suit the needs of the local elite. For example, for many years a reduced number of *departamentos* was maintained in order to neutralize the population increases in the southern part of the province, thus diminishing that region's political influence.

9. The *Partido Constitucional* was a brief experiment (1883–1886) that sought, behind the banners of the defense of universal manhood suffrage, to create a more democratic and participatory party with which to challenge the hegemony of the ruling elite, whose regime was popularly known in the province by several names: *situacionismo*, *iriondismo* (from the name of its principal leader, Simón de Iriondo), and *autonomismo*. To the *Partido Constitucional* flocked breakaway groups from the local ruling elite (the so-called *liberales*), sectors of Rosario's small bourgeoisie, and the *colonos*. The party was unsuccessful in its attempt to establish new rules of the political game and the practices of machine politics prevailed.
10. M. Bonaudo and E. Sonzgoni, "Redes parentales y facciones en la política santafesina, 1850–1900," p. 102.
11. Halperín Donghi, "1880: un nuevo clima de ideas," p. 251. Alberdi made the distinction between the need for "civil liberties" and "political rights" in his prescription for Argentine democracy and state formation.
12. On the 1890 crisis see A. Ford, "La Argentina y la crisis de Baring de 1890," in ed. Marcos Giménez Zapiola, *El régimen oligárquico* (Buenos Aires: Amorrortu, 1975), p. 116.
13. Marcelo Carmagnani, "Las finanzas de tres estados liberales: Argentina, Chile y México," (unpublished), pp. 5–6, 8.
14. Hilda Sábato, "Ciudadanía, participación política y formación de una esfera pública en Buenos Aires, 1850–1880," p. 62.
15. In November 1891, the provincial legislature proposed for the first and "only time," a tax on all sales of wheat and linseed in the province. The statutory decree, passed a month later, confirmed that the commercial sector was responsible for payment of the tax. But a year later, a new law was promulgated that not only codified the law but transferred responsibility for the tax to producers.
16. Bonaudo, Cragnolino, Sonzgoni, p. 263.
17. *La Unión,* 1 November 1891.
18. Bonaudo, Cragnolino, Sonzgoni, "La cuestión ...,"
19. *La Razón,* 5 February 1893.
20. *La Razón*, 11 February 1893.
21. *La Razón,* 8 February 1893.
22. These demands had already appeared among some of the agricultural colonies in the 1870s. See Bonaudo, Cragnolino, and Sonzgoni.
23. The grain tax especially was a provincial tax that the *colonos* felt, as the major payers of such a tax, ought to be administered at the municipal rather than the provincial level. Their argument was that they knew the necessities of the municipalities and could better direct public spending if the revenue remained in local hands. Similarly, by virtue of their access to membership on the municipal town councils, they sought to become the administrators of such tax revenue.
24. *La Razón*, 7 February 1893.

25. *La Razón,* 12 February 1893.
26. E. Cibboti, "La elite italiana de Buenos Aires: el proyecto de nacionalización del '90," in *Anuario* 14; Romolo Gondolfo, "Inmigrantes y política en Argentina: la revolución de 1890 y la campaña en favor de la naturalización automática de residentes extranjeros," in *Estudios Migratórios Latinoamericanos* 17, (CEMLA, Buenos Aires, 1991); Lilia Ana Bertoni, "La naturalización de los extranjeros, 1887–1893: ¿derechos políticos o nacionalidad?," in *Desarrollo Económico* 125 (1992).
27. Bonaudo, Cragnolino, Sonzgoni, p. 263.
28. L.A. Bertoni, p. 71.
29. Bonaudo, Cragnolino, Sonzgoni, pp. 266–267.
30. Bonaudo, Cragnolino, Sonzgoni, p. 268.
31. *La Unión,* 15 November 1891.
32. *La Unión,* 11 January 1894.
33. *El Orden,* 13 July 1897.
34. Electoral Platform of the *Unión Cívica,* 15 January 1891(Rosario) reprinted in *La Nación* (1891), p. 6.
35. W. Landerberger and F. Conte, *Unión Cívica. Su origen, organización y tendencias, 1889– 1o de setiembre-1890, Buenos Aires, 1890,* pp. 351–352.
36. Tulio Halperín Donghi, *Una nación para el desierto argentina* (Buenos Aires: CEAL, 1982), p. 67, 70. "*Mitrismo*" is a term referring to the supporters within the *Partido de la Libertad* of General Bartolomé Mitre, a former president and paladin of Argentine liberalism.
37. W. Landberger and F. Conte, p. 327. Barroetaveña is one of the principal figures in the new party and to whom was entrusted the drafting of the *UC*'s party platform.
38. Halperín Donghi, *Una nación para el desierto argentino,* p. 68.
39. *La Capital,* 27 August 1897. '*Autonismo*' or '*situacionismo*', again, refer to one of the two factions in which provincial liberalism was divided following the overthrow of Rosas. This faction held a hegemony of political power from 1868 onwards.
40. *La Bomba,* 27 May 1894; 29 July 1894; 2 September 1894.
41. *La Bomba,* 29 July 1894: *La Razón,* 31 July 1894.
42. Among the provincial newspapers were *El Municipio, La Razón, La Bomba, La Capital, El Liberal de Rafaela, El Tribuno de Santa Fe,* and others.
43. *Expedientes Civiles,* 8 December 1893 and 9 December 1893.
44. *Expedientes Civiles,* 13 December 1893.
45. *La Bomba,* 1 July 1894.
46. *La Unión,* 29 November 1891.
47. Archivo General de la Nación, Roca papers, Legajo 59, August 25, 1890. Gálvez was an important figure within the *situacionismo* faction, which came to power in the years 1886–1890.
48. *La Razón,* 17 February 1893.

49. In 1891, a breakaway faction of the *Unión Cívica* is established, calling itself the *Unión Cívica Radical (UCR)*. Behind the leadership of Leandro Além, the *UCR* quickly supplants the *UC* as the dominant opposition political force.
50. *Cuaderno Impreso Ilustrado,* 30 July 1893–1894, p. 3.
51. Ezequiel Gallo, *La pampa gringa* (Buenos Aires: Editorial Sudamericana, 1984), p. 377.
52. *La Unión,* 15 November 1891.
53. *L'Operaio Italiano,* 14 September 1893, 17 September 1893; *La Bomba,* 29 April 1894, *La Unión,* 1 November 1894, 30 December 1894.
54. *El Orden,* 7 April 1896.
55. *La Razón,* 31 July 1894.
56. M. Bonaudo and E. Sonzogni, "Redes . . . ," p. 107.
57. Rosario's bourgeoisie does not seem to have been concerned about naturalization because, on the one hand, it felt that the consular institutions of their countries of origin and the networks of ethnic solidarity still were effective and offered them the kind of support they could not expect to find in the national or provincial governments. Moreover, it believed there were other ways to exercise political influence, greater possibilities for collaborating with sectors tied to the government than for allying with the *colonos*. Indeed, such cooperation with the ruling party bears fruit in certain common undertakings (the creation of the *Banco Provincial,* for example). For the moment, Rosario's bourgeoisie is interested in guaranteeing a certain control of the municipal government and exercising hegemony in the business world. Although the constitutional reforms of 1890 momentarily remove electoral possibilities for them, the impact does not seem to be so great among the Rosario bourgeoisie as it is for those who seek naturalization as the solution for political influence. Later, we will see, the Rosario bourgeoisie will seek, through the *Liga del Sur,* to create conditions to contest control of the provincial government. By this point, they had begun to rethink their previous strategy.
58. *El Rosario,* 9 August 1893.
59. *El Rosario,* 7 August 1893.
60. M. Bonaudo and E. Sonzogni, "Redes . . . ," pp. 107 and ss.
61. *El Municipio,* 9 February 1893.
62. *La Unión,* 11 February 1892.
63. *La Bomba,* 13 January 1895; 20 January 1895.
64. *La Unión,* 13 December 1894.
65. *El Rosario,* 21 June 1893.
66. *El Rosario,* 30 June 1893.
67. *El Rosario,* 15 July 1893.
68. *El Rosario,* 22 July 1893.
69. *El Rosario,* 5 July 1893.
70. Municipality of Rosario. List of the individuals who formed part of the "Consejo Deliberante," "Consejo Ejecutor," and the "Comisiones Administradora," from 1873 until the present (Rosario, 1989), pp. 19–21. These individuals were

leading members of Rosario's business community and their exit from municipal politics signaled an important shift in political power to new groups.

71. *La Unión,* 31 January 1895.
72. *La Unión,* 24 January 1895.
73. *La Unión,* 23 January 1896.
74. *La Unión,* 6 February 1896.
75. *La Unión,* 26 January 1896.
76. *La Unión,* 26 January 1896.
77. *El Orden,* 28 February 1896.
78. In this period, the "modernistas" create the *Partido Republicano* and the sectors that come together around Juan B. Justo make their first appearance through the *Partido Socialista.*
79. *El Municipio,* 1 July 1897.
80. *La Capital,* 17 August 1897.
81. *El Orden,* 28 August 1897.
82. *El Orden,* 28 August 1897.
83. As previously explained, *situacionismo* was one of two factions or parties that disputed political power in Santa Fe. The other was the liberal faction. The *situacionistas* (alternately called, the *iriondistas* or *autonomistas*) participated in the so-called *Club del Pueblo*, a rigidly hierarchical political organization that, in the political factionalism of the times, was highly personalist and loyal to the figure of the leader, lacking a program or principles beyond co-opting voters via clientilistic practices. Bernardo de Iturraspe was a leading practitioner of this clientilistic style of politics. A successful businessman, with interests in commerce and the business of agrarian colonization, the latter activity permitted Iturraspe to have fluid contacts with the *colonos* and to be familiar with their needs and concerns. Moreover, he was linked by family ties, thanks to one of his many marriages, to a *colono* family. His distancing himself from the more inflexible, hardline sectors of *situacionismo* and his ties with the *colonos* explain the consensus he achieved in his bid for provincial power.
84. *La Capital,* 7 September 1897 and 21 September 1897.
85. *El Orden,* 12 October 1897.
86. *El Orden* and *La Capital* systematically included the list of supporters, the documents drafted in different areas and by different individuals, as are recounted and recorded in the social and political demonstrations in support of the candidate.
87. Bonaudo and Sonzogni, "Redes . . . ," p. 84.
88. E. Gallo, pp. 426–427.
89. *El Orden,* 17 October 1897.
90. *La Capital,*30 November 1897.
91. *La Capital,* 22 July 1902 and 23 August 1902.
92. *La Capital,* 3 January 1904.
93. *La Capital,* 24 February 1904.
94. *La Capital,* 17 January 1904. In De la Torre's "letter-platform," the political leader threw his full support behind a program of economic development, the

fomenting of education in order train workers and citizens, and the adoption of a economic policy that would increase production while alleviating the burden of taxation.

95. *La Capital,* 17 January 1904.
96. *La Capital,* 15 March 1904.
97. *La Capital,* 3 January 1905.
98. *La Capital,* 11 January 1905.
99. For the case of Córdoba, the subject has been considered by Javier Moyano, "El Comité Electoral Municipal del Comercio. Comerciantes y política municipal en Córdoba, 1908–1909," Córdoba, 1993 (unpublished); and Hernán Ramírez, "Política y grupos de presión. La actividad de La Bolsa de Comercio de Córdoba," Córdoba, 1993 (unpublished).
100. Mansajes de Gobernadores, 1900–1907, Santa Fe.
101. Ibid.
102. *La Capital,* 8 July 1908.
103. *La Capital,* 11 July 1908.
104. *La Capital,* 3 April 1908.
105. *La Capital,* 4 April 1908.
106. *La Capital,* 22 April 1908.
107. *La Capital,* 22 April 1908.
108. *La Capital,* 15 September 1908 and 1 October 1908.
109. *La Capital,* 6 August 1908.
110. *La Capital,* 13 May 1908.
111. *El Orden,* 13 July 1908.
112. *La Capital,* 13 November 1908.
113. *La Capital,* 5 November 1908.
114. *La Capital,* 7 November 1908.
115. *La Capital,* 11 November 1908 and 13 November 1908.
116. *La Capital,* 4 December 1908.
117. *La Capital,* 4 December 1908.
118. Juan Alvarez, *Historia de Rosario (1869–1939)* (Rosario: Universidad Nacional del Litoral, 1980), p. 572.
119. The right to vote was only granted to those who "meet the qualifications of having resided in the country during a prescribed period, one established by the law, and that they be owners of real estate or, the second prerequisite being absent, that they be fathers of Argentine children." See Enrique Thedy, "Indole y propósito de la Liga del Sur," in *Revista Argentina de Ciencias Políticas* 1, Buenos Aires (1910): 91.
120. E. Thedy, pp. 89–90.
121. *La Capital,* 4 December 1908.
122. *La Capital,* 31 October 1908.
123. *La Capital,* 3 January 1909.
124. Juan Alvarez, p. 576.

CHAPTER TWO

Water, Guns, and Money: The Art of Political Persuasion in Mendoza (1890–1912)

Joan Supplee

Introduction

By 1890 Argentine politicians followed well-established patterns. The elites who managed national and provincial governments inherited from their predecessors systems of power and patronage and techniques of electoral persuasion. Individuals entered government service to enhance family fortunes and position. A family's success in serving both clients and patrons shaped its political and economic fortunes. After independence, the native elite broke into competing factions on both the national and provincial levels. While maintaining their traditional view of power and patronage, the elite parties developed new techniques and adapted old ones—influence peddling, vote tampering, intimidation, and violence—to ensure their power within a new electoral system. As long as economic opportunity and access to wealth remained restricted, these methods effectively limited access to political power.[1]

The traditional elite in Mendoza defined the boundaries of its political power and patronage within the broader context of national politics. From independence until 1880, provincial leaders fought with and against national leaders over the structure of the republic. Between armed confrontations, local leaders controlled their provinces and honed their skills at electoral manipulation. In Mendoza, local *caudillos* based their power on control of the province's ranching economy and local government office.

Families such as the Villanuevas, Civits, Seguras, and Ortegas dominated the province. After national consolidation, provincial leaders interacted with the national oligarchy, known the Generation of 1880, to define and defend their power. As long as tranquility and the national ruling party, *Partido Autonomista Nacional* or *PAN*, prevailed, the Generation of 1880 collaborated with provincial political machines. Local leaders skillfully manipulated the electoral system within their boundaries unchecked by national authorities.[2] As one Córdoba newspaper, *La Carcajada*, described the system, "Elections in the Argentine Republic! Ha! What a joke!... Today the Argentine Republic is a fiefdom. The will of the masters is what prevails."[3]

This chapter will analyze how the traditional political system broke down in one province, Mendoza (see map III). Mendoza's western location and its unique economic development allowed it a greater degree of autonomy than provinces closer to the Atlantic coast. Despite this independence, the breakdown of elite hegemony in the province in many ways was closely linked to the failure of the elite at the national level. In 1890, the traditional elite at both levels faced the first of a series of political and economic crises that tested their power. They survived, but economic changes steadily eroded the basis of their power. Mendoza's transition from oligarchic to mass-based politics reveals the degree to which this was a contested process characterized by prolonged conservative opposition. After 1902, the Mendozan elite resorted increasingly to repressive methods of control with the period of most extreme political repression occurring between 1906–1910. Fraud and armed intimidation in the political arena jeopardized the very legitimacy of the political system. National and provincial reforms in 1912 saved the system from collapsing, but marked the decline of oligarchic rule in both the province and the nation. Liberal historiography, Argentina's version of Whig history, conveys a sense of peaceful accommodation, codified in the 1912 'Sáenz Peña Law,' and underestimates when not ignoring outright the degree to which this was a conflictive process whose outcome was by no means preordained. Mendoza's history thus illuminates issues long ignored by political history regarding the origins of the "democracia de masas" in twentieth-century Argentina.

Elite Politics

For 50 years after the *cabildo abierto* met in the Plaza de Mayo in Buenos Aires and declared Argentina's independence, Liberals and Conservatives within the landed elite disputed and divided among themselves the spoils of local provincial government. The victory of the Liberals at both the local and national level in the 1860s seemed to put an end to these regular but discrete swings in local political fortune.[4] Unchallenged, the Liberals then instituted their plan for economic modernization. National economic changes, however, undermined the social basis of the oligarchic system. Expansion of the export sector of the economy combined with an influx of immigrant labor and foreign capital to fuel development of the internal market and local infrastructure. New economic pursuits gave rise to political interest groups whose members were outside the elite-based system of power and patronage and who were immune to traditional methods of social control.[5] To retain its power, the elite directed more government resources into electoral control. Continued economic expansion exacerbated conflicts between the traditional elite and its new opponents.

Changes in the national economy had a profound effect on Mendoza. Vine cultivation and wine production replaced ranching as the mainstay of the local economy. Italian and Spanish immigrants lured to the province by the promise of land and good wages supplanted *criollo* and Chilean ranch hands as the primary labor force. In short order, many immigrants became landowners and merchants, allied themselves with native *mendocinos* on the basis of economic rather than familial ties, and disrupted established systems of status, wealth, and patronage.[6]

The first serious fissures in the traditional political order appeared within the Mendozan ruling class in the late 1880s. Landholders who had switched from ranching to vine cultivation clashed with ranchers over the credit policies of the provincial bank. Once in power, the viticulturists used provincial government resources as they had always been used—to serve the interests of the ruling party and its members. In January 1889 a powerful rancher, former governor and national senator from Mendoza, Colonel Rufino Ortega, took advantage of temporary divisions within the national party and his command of the 12th Regiment stationed in Mendoza to oust the elected government of new viticulturist Tiburcio Benegas. His coup brought Mendoza to the forefront of national concerns and forced national leaders to intervene. Acting president Carlos Pellegrini declared that he would not permit anyone to arrive "at the sad conclusion that one can change the political system with an act of audacity or force." Pellegrini sent Senator Manuel Derqui as a mediator to restore legitimate government. At the same time, President Miguel Juárez Celman, on leave in Córdoba, sent Federal Justice Calixto Torres to investigate the matter. Such disputes in Mendoza were over economic policy and political succession, paralleling a similar schism in the national leadership. After a few days in the province, Senator

Derqui insisted on the restoration of Benegas as governor and ordered Ortega to return all weapons used in the coup to the national armory. The colonel was also transferred to a post outside the province. Despite raising arms against the provincial government, Ortega soon reclaimed his Senate seat.[7] His status within the local elite and links to national leaders saved him from meaningful punishment. This incident strained, but did not break the patronage system. Future power plays would not be so amicably resolved.

The collapse of the financial system in 1890 fragmented national and provincial systems of power and patronage. As control of the provincial government and its resources became essential for economic success, irreconcilable differences developed within the ruling class. The new governor, Oseas Guiñazú, changed political alignments five times during his 25 months in office. He was so busy "executing numerous and undignified pirouettes in the governor's seat,"[8] according to one pundit, that he had time for little else. The economic crisis split the ruling class into three separate parties: the Liberals, the Radicals, and the Intransigent Radicals. New to the national scene, the Radical Party challenged the *PAN*'s political monopoly and criticized its economic policies. In Mendoza, the political fragmentation finally drove Guiñazú to resign in October 1891. An agreement between former presidents Julio A. Roca and Bartolomé Mitre brought some measure of local order as Ortega partisans, a section of Radicals loyal to Mitre, and the Liberal Party came together under the leadership of Emilio Civit. Scion of a powerful political family, he was linked by marriage to two others: the Ortegas and the Benegas. His political allies also included members of the Villanueva family and national leader Julio A. Roca.

Despite this new national and provincial alliance, political tensions did not abate. Guiñazú's replacement as provisional governor, Pedro Ortiz, belonged to the Radical Party. Once in office, however, he betrayed the Liberal-Radical-Ortega alliance by switching loyalties to a Radical faction controlled by rancher José Nestor Lencinas. Ortiz's defection to Lencinas split the Radical Party so irrevocably that not even a visit from national party leader Leandro Além could heal the breach. Under Lencinas's guidance, Ortiz tampered with voter registrations, attempting to purge the lists of all but Lencinas supporters. He also appointed Lencinas head of the city government. In January 1892, before Ortiz could use his office to secure an electoral victory for his faction, deputies in the provincial legislature loyal to Civit wrote to the Minister of Interior, José M. Zapata (a *mendocino* whose vacated national senate seat went to Emilio Civit) requesting federal intervention to preserve the national constitution and protect liberty in Mendoza. Radical deputies responded by blocking legislative action; Civit's deputies threatened to suspend Governor Ortiz. Zapata ordered Ortiz to work with the dissatisfied deputies or risk federal action. In the interim, Radical deputies launched an armed attack against their legislative opponents to stop them from removing Ortiz. Several deputies were wounded in the melee and the Chamber's secretary was killed. It was enough to bring national intervention and prevent a Radical electoral victory. On January 22, Zapata appointed Francisco Uriburu as interventor and declared martial law. Uriburu removed Radicals, including Lencinas, from the Mendoza city government and electoral boards throughout the province. To isolate further the Lencinas faction, Francisco Civit (a former governor himself) advised his son to either become a Radical or ally with them in order to maintain power. The union between the *orteguistas*, Liberals, and Radicals was formalized in February 1892 with the creation of the United Parties (*Partidos Unidos* or *UP*). Elections carried out under the intervention brought victory to the new party. The *UP*'s connection with the *PAN* helped it secure control over the provincial government in 1892.

In response to their defeat, the Lencinas Radicals pledged to boycott future elections and seek new avenues of support. Their inability to work with the ruling coalition cut them off from the traditional circles of power. To compensate, they courted new members of Mendozan society beyond the networks of power and patronage controlled by the UP—immigrant laborers and entrepreneurs.[9] Unwittingly, the *UP* had promoted the creation of a new and potentially formidable opposition.

With the disaffection of Lencinas, the *UP* became the bulwark of the traditional elite and its patronage system. The *eminence gris* of the party was Emilio Civit. Under his leadership, the *UP* fine-tuned traditional methods of electoral control.[10] What it could not obtain through patronage, it won through fraud. As one opposition paper described the party:

> *Civitismo*, a family oligarchy, has been pursuing public office since 1861, using all means to maintain itself in positions of power. When it believes that it is losing, it seeks to rectify its losses by changing political alliances and betraying its friends. It has no political ethics. The secret of its apparent strength is political betrayal—then, now and later.[11]

Civitismo continued to dominate provincial elections until 1912, but Civit preferred to exercise power indirectly, pulling strings from behind the scenes. He spent most of his time in the national capital, but his influence was widely felt in the province. In 1894, Governor Pedro Anzorena, a Radical who had joined the ruling coalition, convened a convention to rewrite the provincial constitution. Emilio's father, Francisco Civit, presided. Anzorena disapproved of the new draft because of the blatant favoritism shown ruling families. The new constitution gave established landowners greater control over water allocation in the province—a power that was critical in determining economic success in agriculture. Anzorena chose to resign in 1894 rather than sign the document or approve a huge concession of land to General Ortega for his services to the province. Most legislators favored replacing Anzorena with Ezequiel Tabanera, Jr., another moderate Radical. Civit favored mainstream *UP* candidate, Francisco Moyano, but realized that he was one vote short of victory. On the day of the election, as Tabanera—already dressed in a tuxedo and gubernatorial sash—awaited word of his election, Civit forged a telegram to legislator Pascual Suárez from Bernardo de Irigoyen, leader of the Radical party. The telegram instructed Suárez to switch his vote to Moyano "for reasons of the highest national political order." The ruse worked. Tabanera was publicly humiliated; the opposition press dubbed Civit "the great elector." Moyano bestowed on Civit the post of minister of finance as a reward for his campaign service.[12] The Tabanera family never reconciled with Civit and moved into the opposition camp.

The *UP* controlled the province with little challenge from 1895 until 1902. The party allocated public resources for improvement of irrigation networks that benefited private lands. When finances fell short, the government under *UP* tutelage printed its own money—indirectly transferring resources from the poorer to the wealthier sectors of the province by inflating the currency.[13] These policies victimized new immigrant landowners and urban laborers who represented a small and powerless sector of Mendozan society in 1895. As the *UP* expanded the agricultural frontier, it simultaneously encouraged the growth of these two groups. The grape harvest attracted laborers to the fields, as did construction on roads, railroads, and canals. In 1893, day laborers received one to two pesos a day for their work and the situation remained roughly the same until 1902. Workers resented the inflation sponsored by the *UP* and looked to other political leaders to champion their cause. The Intransigent Radicals, led by Lencinas, openly courted workers during this period.[14] But as long as the economy remained strong, the *UP* had nothing to fear from workers or Intransigent Radicals.

An economic crisis in 1901–1902 provided the first indication that the *UP*'s grip on Mendoza was slipping. Collapse of the market for provincial notes caused workers to riot in the capital in 1902, and led to the *UP*'s first electoral loss. The weakened economy exacerbated divisions below the surface of provincial politics. Alienated members of the governing elite made common cause with small farmers in the southern department of San Rafael who were dissatisfied with *UP* distribution of public funds. Independent Party provincial deputy candidates Juan E. Serú, a grape grower, merchant, and former *UP* member, and Abelardo Tabanera, a politically-active landowner with a family ax to grind over the 1894 humiliation, defeated *UP* candidates in March 1902. Their victory caught the *UP* off guard and the governor failed in his attempt to overturn the results.[15] The economic crisis thus served to unite alienated members of the traditional elite with a new force—small immigrant landowners—and forged the first successful challenge to the political hegemony of the *UP*.

UP Governor Elías Villanueva (1901–1904) responded immediately to this challenge, but did not anticipate the changes that would accompany his actions. First, he revised public works policies and water grants in the San Rafael district. He conceded water to large parcels of land in the region held by politically influential individuals and companies. These concessions inflated land values and bought the *UP* goodwill among large landowners, but did not slow the erosion of party adherence in the south. One particular grant highlighted the problem. In January 1902, the legislature approved a water concession from the Atuel River sufficient to irrigate 10,000 hectares. The grantees, Mario and Pío Perrone, proposed to establish an agricultural colony in the Colonel Beltrán section of San Rafael. According to provisions of the concession, the Perrones had five years to settle and cultivate the land or forfeit their water rights. A privately constructed canal carried water to the fields in 1908 and the Perrones received permanent water rights in 1910. Rather than allying themselves with the *UP*, however, the Perrones, Italian immigrants, established links with members of the old guard hostile to the Villanueva regime. To build their canal, they hired civil engineer Exequiel Tabanera, relative of the recently elected deputy, Abelardo Tabanera. The Perrones also subdivided and sold their land to immigrant agriculturists who used private rescues to build the irrigation infrastructure.[16] The *UP* counted on large landowners and companies to maintain its influence in the district, but by 1902 and increasingly thereafter, the district was populated by small, immigrant landholders who owed no familial, financial, or occupational allegiance to the local elite.

Because the system of power and patronage had broken down in the south, Villanueva's successor, his nephew Carlos Galigniana Segura, faced a new electoral challenge in San Rafael in 1904. He attempted to reestablish *UP* dominance with a carefully orchestrated provincial senate campaign. The official candidate was Enrique Day, the outgoing vice-governor. In spite of widespread vote tampering and police intimidation on behalf of the *UP* candidate, Independent Party candidate Ezequiel Tabanera, Jr. carried the day. Thwarted at the polls, the governor had the election invalidated by presenting evidence of electoral fraud to the legislature—fraud that had been authorized by the governor himself.[17]

Immediately following the election, Galigniana Segura moved to convince San Rafael voters of his party's concern for their welfare. He sent the director of water management on an inspection tour of the district canals to assess needed improvements. Before calling for a new election for provincial senator, the governor authorized the minister of government to allocate funds for reinforcing flood walls on one of the main rivers. He also disbursed funds to his administrative head (*jefe político*) for water improvements. A leading national newspaper, *La Prensa*, reported that Galigniana Segura's real purpose was to buy votes, not build canals.[18]

The United Parties resorted to more direct methods of persuasion when these efforts failed to sway voters. Emilio Civit, then serving as national minister of public works, threatened to cancel all public works in San Rafael if voters persisted in their obstinate support of Tabanera. A faction of landowners loyal to the ruling party managed to move the departmental capital from San Rafael, the seat of resistance to the *UP*, to Colonia Francesa, where they could control elections. They also arranged for the railroad to bypass San Rafael in favor of Colonia Francesa. These combined efforts secured the election for the *UP*'s candidate, José Salas, in 1904.[19]

Elite Politics under Siege

The electoral triumph proved short-lived. Troubled by the electoral fraud and ensuing public outcry, the *UP*'s candidate resigned and his seat remained vacant until 1906. The obvious political corruption and lack of representation in the provincial capital encouraged foreign residents in the area to register to vote (citizenship was not a requirement for provincial elections) as a way of resisting *UP* dominance. They were successful. Despite the *UP*'s best efforts, it lost control of the San Rafael city government at the end of 1904.[20]

To recoup its losses and to prevent such defiance from spreading, the *UP* administration shifted public funds into development of a provincial police force. In 1904 a new rural constabulary force was created. It consisted of a mobile company of over 300 men with jurisdiction throughout the province. The governor also created a security squadron of 26 men to assist police in the provincial capital. The squadron acted as an elite corps of spies to harass members of the opposition. The size of the rural police force was increased by ten percent in 1906.[21] To win elections after 1904, the *UP* increasingly relied on these new armed forces to limit voter registration, intimidate opposition members, and secure favorable outcomes at the polls.

This strategy strained provincial resources. To meet budget shortfalls, the government increased water taxes, printed more money, sold public land, and cut back on provincial services. A new business newspaper controlled by a faction of the Radical party, *El Comercio*, decried increased water taxes, as high as four pesos per hectare in some departments, and dramatically charged that increased burdens were forcing many northern irrigators into penury. The governor's fiscal policy also antagonized public employees, traditional clients of the *UP*. When Villanueva left office, the provincial government owed provincial teachers four months' back pay. Villanueva's budget devoted less than 10 percent to education while allocating nearly 30 percent to public safety. Without new sources of revenue, landowners, laborers, and public employees shouldered the costs of political repression.[22] The result was a constant drain on *UP* strength as supporters and clients found the opposition more sympathetic to their economic plight.

The 1906 gubernatorial campaign revealed an increasingly volatile political atmosphere. Emilio Civit took his party's nomination for governor. His candidacy was controversial and precipitated a split in the party; some members joined a coalition of opposition parties in an attempt to prevent his election. The election surpassed all previous ones in terms of fraud and repression. The government of Galigniana Segura transformed the governor's palace into election headquarters. Litigants with suits pending in provincial courts were informed that if they did not support the *UP* the outcomes of their cases would not be favorable. The party used its control over voter registration to ensure a *UP* majority. Names of supporters were added to multiple department registers (some of them deceased) while in areas where rivals held a clear majority registration offices never opened. One opposition newspaper alleged that government loans earmarked for public works financed the *UP*'s electoral campaign. At the same time, the government bolstered police forces, and brought in recruits from neighboring provinces who promptly registered to vote in Mendoza. The weapons—stored in the province after the Radical revolt of 1905—used by these new recruits belonged to the national government. A new squadron of 50 cavalry was stationed near the capital and an additional 50 men were added to the force already in San Rafael. Opponents charged that Galigniana Segura and Civit had turned the province into "an armed camp."[23] Local party leaders also harassed members of the opposition, threatening to cut off their water allocation and to interfere with their businesses.[24] The Electoral Commission objected to the atmosphere of violence and the absence of constitutional guarantees and called on the federal government to nullify the vote. President José Figueroa Alcorta, alarmed by public disturbances and reports that the provincial government was amassing arms to repel a rumored invasion from the province of San Luis, sent Leopoldo Basavibaso to assess the need for federal intervention. Basavibaso persuaded Governor Galigniana Segura to return weapons taken from the national armory, but could not prevent him from releasing 50,000 pesos for the purchase of weapons for his prison guards. Opposition members presented the president's representative with evidence of massive registration fraud: of Mendoza's 40,000 eligible voters only 17,500 had been registered. Despairing of a peaceful outcome, Basavibaso recommended federal intervention.[25]

The threat of federal intervention loomed as Emilio Civit assumed the office of governor in early 1907. Relying on political savvy acquired during his service as national minister of public works, he dissuaded Figueroa Alcorta from intervening in the province. In exchange, the governor assured the president that he would stay out of national politics.[26] Both remained true to their bargain.

Having secured himself a free hand in the province, Civit turned his full attention to the challenge presented by his opponents, especially party members who had left the *UP* to form the Civic Party. He stretched public credit to the limit as he pursued the dual goals of promoting economic development and preserving political control. He spent heavily on both weapons and irrigation expansion—more than any previous administration—to punish his enemies and reward his followers.

Although Civit was deeply committed to a program of state-directed economic development, he was no less committed to *UP* control of that development. He proposed legislation to complete the government's control of water distribution. Opponents feared that the new legislation gave too much discretionary power to the governor. The Civic Party, for example, attacked recent water grants as damaging to Mendoza's heartland. Leading opposition newspapers urged legislators to reject the governor's plan. The campaign against the bill was successful enough to delay the measure in committee. Civit countered by offering political favors and constitutional reform. Knowing that they lacked a majority in the lower house to get Civit's bill passed, *UP* delegates called a special session of the Chamber without informing opposition delegates. Civit then leaned heavily on recalcitrant senators to approve his package in 1908.[27] The *UP*'s loss of its legislative dominance and its reliance on force, bribes, intimidation, and chicanery overshadowed the governor's victory.

Civit's absolute control over water distribution allowed him to victimize unrepentant members of the electoral coalition that had opposed him. His plan to "rationalize" water control in the province included strong regulations against water fraud. It gave him power to levy stiff fines subject to arbitrary increases. In some cases these fines equaled the value of the land holdings against which they were levied. To root out violators, the law provided generous remuneration for informers. Two opposition papers soon discovered that a certain "Manuel O. Hidalgo" had made an extraordinary number of denunciations in 1907 and 1908. The mysterious Hidalgo surreptitiously covered large sections of the province detecting violations by outspoken critics of the Civit regime, including former governor Pedro Ortiz and Exequiel Tabanera. Clandestine water use by party members was tactfully overlooked. The civic-minded Hidalgo never appeared to collect the levied fines, fueling the belief that he did not exist. In 1911, *La Tarde* reported that a former Civit cabinet member and an engineer in the water management office had accused each other of being "Señor Hidalgo."[28] As opponents had feared, Civit gained a powerful tool with his new water legislation. It became a subtle and cost efficient way to harass political enemies.

The critical flaw in the water management system stemmed from the governor's inability to separate development plans from political ends. Civit clearly supported water projects that profited supporters. In 1909, he approved his friend Juan Franco's proposal for construction of a public slaughterhouse that involved an enormous concession of water. Civit also sanctioned a 600,000 peso irrigation project that directly benefited his own property and that of his cronies. Even more flagrant were the water grants in the southern part of the province. Management of southern water increased in political value as more members of the ruling party invested in San Rafael real estate. This trend accelerated after the rail line reached the area in 1903 and brought the onset of public land auctions. Galigniana Segura and his chief of provincial police, Octavio Fernández, purchased the largest tracks of land at a May 1907 auction. Deputy Lucio Funes, armed with complaints about insufficient water supplies in San Rafael, managed to block the governor's request. The setback proved temporary. The 1908 Water law gave the governor and his friends new access to water in the south. Of the major concessions made in 1908, nearly 90 percent went to supporters of the governor (see Table 2.1). The most controversial came in April 1908. In another unscheduled meeting of the legislature, Civit's allies approved Alfredo Israel's petition for water to irrigate 90,000 hectares in the

Table 2.1 1908 Water Concessions Made by Emilio Civit

Grantee	*Area in hectares*	*Department*	*Source*
Alfredo Israel*	90,000	San Rafael	Diamante
Carlos Gónzalez and Ventura Segura*	2,012	Santa Rosa	Tunuyán
Carlos Galigniana Segura*	1,886	Santa Rosa	Tunuyán
Juan E. Surú*	1,000	San Rafael	Diamante
Luis V. Spineto*	5,000	San Rafael	Diamante
Jacinto and AugustínAlvarez*	1,500	San Rafael	Diamante
Engelber, Hardt y Cía.	500	San Rafael	Diamante
Rosario Martínez	2	Maipú	Canal Bovedas (Mendoza)
Victor Manuel Zuloaga	17	Las Heras	Hijuela Segura (Mendoza)
Modesto Ignacio Colori and Domingo Varela	1	Guaymallén	Hijuela Tobar (Mendoza)
Cipriano C. Ibáñez	100	Lavalle	Hja Arancibia (Mendoza)
Abelardo Capdevilla*	333	Las Heras	Hja. Algarrobal (Mendoza)
Carlos Cánepa	100	San Martín	Hja. Lucero (Mendoza)
Juan Franco*	4,000	San Rafael	Atuel
Orlando Williams and Carlos Shaw	12,000	San Martín	Mendoza
Carlos González*	100	Santa Rosa	Canal Dormida (Tunuyán)
Alfredo Ahumada*	36	La Paz	C. M. La Paz (Tunuyán)

Total Hectares Granted: 118,087
Total Hectares with * (Indicates grantee with political ties): 105,534
Source: Mendoza province, *Registro oficial de la provincia de Mendoza 1908* (Mendoza, 1908)

Monte Coman section of San Rafael. In 1910, *Los Andes* exposed Civit as a silent partner in the deal.[29] He routinely sacrificed rationality in water distribution to the exigencies of party politics, using water rights to reward his followers and punish his enemies among the *criollo* elite.

When the system of water-based patronage failed to achieve his ends, Civit resorted to arms and intimidation to silence rivals. He devoted nearly half of the provincial budget to enlarging his security forces. In 1907 he hired 200 additional police and then raised their salaries. The new recruits brought the number of men under arms to 800. With 500,000 pesos

Galigniana Segura had allotted for weapons, Civit ordered 1,000 Winchester rifles that were later confiscated by the federal government. Civit defended his military buildup by citing political instability in neighboring San Juan and San Luis provinces and plots against his government. On two occasions, the government staged "rebellions" against itself to justify the arms buildup. The force that Civit created, however, was more effective against electoral than bogus revolutionary challenges to his regime. Opposition deputies, alarmed by the number of Civit's troops, attempted to reassign some of them to the fire department, but failed. Civit further supplemented his armed force by creating a troop of mounted guards to defend outlying areas against bandits. Both *La Prensa* and *Los Andes* reported incidences of troops used to support official candidates or to keep opposition voters from the polls.[30]

Part of Civit's corps functioned as a provincial spy network, keeping him informed of the opposition's movements. In general, the surveillance teams did not hide their activities. Two spies followed Ezequiel Tabanera, Jr. when he traveled to Buenos Aires to meet with the president. An agent that closely resembled aging Radical Pedro N. Ortiz and dressed just like him—down to the top hat and cane—was stationed outside of Ortiz's home. He tagged Ortiz everywhere, so intimidating the old man that eventually he refused to leave his house. Later, Civit's agents tried to link Ortiz to one of the staged coups against the government. More troublesome than this type of harassment, leaders of the opposition often had their homes searched on the basis of blank search warrants. At the same time, Civit established a team of "wine police." Inspectors routinely delayed wine shipments and levied heavy fines on wineries owned by opponents of the regime. The establishment of an independent inspection office by the Wine Exchange reinforced charges by victims of Civit's "wine police" that the force was politically directed. The Governor also carried his political vendettas to the second generation. He denied a provincial scholarship for the study of oenology to Enrique Silvetti, a qualified candidate, because his father was prominent in the Radical Party and an outspoken opponent of the *UP*.[31] While his regular police harassed political enemies on the streets of the provincial capital, his wine police threatened their businesses.

The limits of Civit's carrot-and-stick policies were clear by 1909—the provincial budget could not support both (see Table 2.2). Too many new players in the political arena were outside the *UP*'s patronage networks and were easily drawn into the opposition. Members of the *UP* had been left behind by the economic changes that overtook Mendoza and provincial finances could no longer bridge the gap between their aspirations and their

Table 2.2 Provincial Budgets and Percentage Apportioned for Security 1902–1914

Year	1902–1914		
	Total Budget in Pesos	Police in Pesos	Percent of Budget
1902	1,712,107	495,348	28.9
1903	1,429,797	465,499	32.5
1904	1,372,675	419,210	30.5
1905	1,684,358	508,353	30.2
1906	1,910,487	657,300	34.3
1907	2,244,790	995,550	39.5
1908	3,343,050	1,281,660	38.3
1909	3,808,275	1,594,762	41.9
1910	4,404,162	2,003,347	45.5
1911	5,084,441	2,108,690	41.5
1912	5,720,035	2,129,740	37.2
1913	6,366,987	2,232,340	35.1
1914	7,074,708	2,330,960	32.9

Source: Archivo Histórico y Administrativo de la Provincia de Mendoza, Carpeta 2 de la biblioteca, "Gastos y Presupuestos" Años 1902–1914.

personal resources. More and more members of the ruling party were excluded from power and access to resources, and abandoned the party for the opposition. Only heavy-handed political manipulation and the police force sustained the *UP*'s hold on the government after 1910.

Civit planned to continue his dominance in Mendoza by nominating his son-in-law, Rufino Ortega, Jr., as the party's standard bearer. He picked the young Ortega to stop his father, General Ortega, from running for office and because he thought he could control his young son-in-law. In this he miscalculated. Provincial credit was dangerously overextended by the time Ortega assumed power and he had to make a hard choice concerning allocation of provincial resources. His priorities drove a wedge into the United Parties and left him outside the faction controlled by Emilio Civit.

Water played a central role in the province's political realignments. A new provincial constitution written before Civit left office gave Ortega greater control over water concessions and enhanced water concessions as a patronage tool. Trouble began in September 1910. A severe drought galvanized residents along the Mendoza River to complain that new upstream concessions made by Civit had deprived them of their guaranteed share of water. The irrigators' plight quickly attracted the opposition's attention and focused on a 1909 water concession made to Juan Franco as part of his slaughterhouse contract with the provincial government. The contract allowed Franco enough water to irrigate 6,000 hectares, all upstream from the central irrigation zone. Opponents alleged the contract allowed Franco to take one-fifth of the river's flow during the dry season. The *UP* hierarchy insisted Ortega honor the Franco grant even though it hurt some party members.[32] By taking a hard line, the party created its own crisis.

Reluctant to move against members of his own party, Ortega refused to endorse either side. He did meet with members of the opposition who presented a strong case against Franco. In response, Galigniana Segura, president of the United Parties, resigned and revealed publicly that he felt betrayed by Ortega. Manuel Lemos, minister of industry and public works, also quit Ortega's administration. Increasingly isolated politically, Ortega modified but did not cancel the concession outright.[33]

The crisis forced Ortega to reassess his political position in Mendoza. He moved to recoup his support in the *UP* by approving a record number of water grants to San Rafael party members. He also turned to cultivating new political allies outside the party, particularly small landowners in the north. He sped up processing water petitions and decreed that adjacent property holders could apply for rights to water running off land with permanent concessions. This regulation appealed to new northern landowners where subdivisions of older parcels led to intensive cultivation of vineyards. Ortega also used provincial funds to improve northern canals. These initiatives benefited small, mainly immigrant property holders, and allowed Ortega to build independent support outside the *UP*.[34]

Civit's plan for southern irrigation drove the final wedge between old guard members of the United Parties and General Ortega. When the local press began investigating provincial finance during the Civit regime, the details of the former governor's extravagant (32 million pesos) plan for southern irrigation became public. *Los Andes* opened its exposé with an examination of the 1909 French loan Civit had negotiated to finance southern irrigation construction. A Civit crony and southern property owner, acting as an intermediary, had pocketed 1.2 million francs for his services. Civit's plan for the loan was to divert water from the Atuel River into the Diamante River across land owned by Civit and his business associates. Landowners downstream protested. As Ortega tried to contain the scandal, *Los Andes* revealed that Civit had signed a 20-million-peso contract with a French firm pending legislative approval.[35] Civit counted on Ortega to deliver the votes to seal the deal.

The scandal forced Ortega to move carefully between the two sides. Under pressure from the opposition and outraged southern landowners, he modified Civit's plan in April 1911 to serve all landowners between the

Diamante and Atuel. He canceled fines pending from Civit's regime against opposition members for alleged clandestine water use. He also reformed the wine inspection process and eliminated many inspection employees who were loyal to Civit. In an attempt to mollify *UP* members, he committed public funds for canal construction to serve their properties.[36]

Ortega's handling of the scandal dismayed Civit and his branch of the *UP*. Civit usurped Ortega's position as party leader and Ortega's own vice-governor, a *civitista*, turned against him. When he presented his slate of candidates for the May provincial elections, party leaders, in an unprecedented move, rejected them and drew up their own list. The split in the party allowed Independent Party candidate Lucio Funes to capture a seat in the Chamber of Deputies. *UP* legislators rejected Funes credentials and nullified the election, but the split between Ortega and Civit weakened the party.[37] It also drove Ortega to court the opposition as an alternative base of support.

As the new year began, diverse sectors of the opposition against the United Parties found common ground. The largest parties—the Independent Party (*Partido Independiente*) and the Civic League (*la Liga Cívica*)—formed a new party, the Popular Party (*Partido Popular*). They united to capitalize on national electoral reforms promised by President Roque Sáenz Peña. Enacted in January 1912, the reform law set new standards for voter registration and instituted the secret ballot. Ortega pledged to uphold the national law.[38]

The provincial elections held in March 1912, demonstrated how effective the opposition could be with the protection of national electoral reforms. Ortega, realizing the strength of the new Popular Party and thwarted in the *UP*, moved to distance himself from his old party. He canceled Franco's slaughterhouse contract just prior to the election. When the Popular Party captured a majority of seats in the legislature, the governor sought their support with an offer of two cabinet positions. Pedro Benegas, Popular Party candidate for national deputy, also beat the *UP*'s nominee, Estanislao Gaviola, a member of the Radical Party. In 1913, the Popular Party gained one of Mendoza's national senate seats with the election of Benito Villanueva. Ortega had supported Villanueva and, in the wake of his victory, canceled Civit's contract with the French company for irrigation work in San Rafael.[39]

A weakened economy further undermined the *UP*'s power. The demand for Mendozan wine slipped in 1912 and continued falling in 1913. Large wine producers, allied with the *UP*, attempted to offset the decline in profits by cutting grape prices. Smaller producers and growers protested and, after Ortega's administration defended the price cuts, turned to the opposition. The recession of 1913 delivered the coup d'grace. An international contraction in credit coupled with a growing budget deficit restricted the *UP*'s ability to serve the economic interests of the larger producers, thereby losing their loyalty. By the gubernatorial election of 1913, the *UP* had ceased to exist as an effective political entity.[40] In November 1913, the Popular Party launched its race for the governorship with the first modern campaign in the province. Party leaders organized and campaigned throughout the province to mobilize the province's middle class voters. Their efforts paid off as the Popular Party candidate, Francisco Alvarez, won the election.[41]

The Mendozan Popular Party was the first to benefit from national electoral reform, but it was not the only opposition party in the province. José Lencinas's Instransigent Radicals reentered the political arena in 1912. More broadly based than the Popular Party, Lencinas' Radical Civic Union (*Unión Cívica Radical* or *UCR*) was based on Mendoza's rural and urban wage laborers. This base allowed it to benefit from national electoral reform and to monopolize provincial government after 1917.[42] No less a political machine than the *UP*, the *UCR*'s leadership was drawn from the same class, but unlike the *UP*, its leaders recognized and tapped the strength of rising economic groups within the province. In many ways, Lencinas's organization, drawing strength from Mendoza's new working class, presaged Juan Perón's political movements by several decades. Both created new systems of power and patronage that, at least initially, did not rely on intimidation, bribery, or violence to achieve electoral majority.

Conclusion

The practice of politics in Argentina during the period of the belle époque was based on concepts of power and patronage and on electoral manipulation that served the governing elite's interests. The ruling class at the national and provincial level used its monopoly of government resources to initiate an ambitious program of economic growth. Its goal was growth without political changes. The result, however, was an alteration in the distribution of both economic and political power that undermined traditional systems of patronage. The case of Mendoza clearly demonstrates the autonomy and strength of provincial elites and the penetrating force of the export-driven model of economic growth that affected both national and provincial leaders. Mendoza's elite redirected the provincial economy using local resources while intending to retain both economic and political control. Forced to choose between the two after 1902, the *UP* selected political power over economic growth. But the changes sponsored by Mendoza's traditional rulers also tied the province to the Argentine economy and the national political structure. When national leaders opted for economic growth over political power and instituted national electoral reforms, they altered the political equation in the province in favor of new economic groups. They disrupted the politics of force, bribery, and fraud. Given the chance to participate, Mendoza's newly enfranchised voters joined with disaffected elites to dismantle the *UP*'s political monopoly and create alternative systems of power and patronage. After 1912, control of water, guns, and money no longer guaranteed electoral success for any party in Mendoza.

Notes

1. For example see: James R. Scobie, *Argentina: A City and A Nation*, second edition (New York: Oxford University Press, 1971); Karen Remmer, *Party Competition in Argentina and Chile, Political Recruitment and Public Policy, 1890–1930* (Lincoln: University of Nebraska Press, 1984); David Rock, *Argentina, 1516–1982* (Berkeley: University of California Press, 1985); Carlos Melo R., "Los partidos politicos argentinos entre 1862 y 1930," in *Academia Nacional de la Historia, Historia Argentina Contemporánea* vol II: 82–91; José Luis Imaz, *Los que mandan* (Buenos Aires: Editorial Universitaria de Buenos Aires, 1964); José Luis Romero, *Las ideas politícas en Argentina* (Buenos Aires: Fondo de Cultura Económica, 1946); Jonathan C. Brown, "The Bondage of Old Habits in Nineteenth Century Argentina," *Latin American Research Review* 21:2 (1986): 3–31; Richard Graham, *Patronage and Politics in Nineteenth-Century Brazil* (Stanford: Stanford University Press, 1990).
2. Lucio Funes, Gobernadores de Mendoza, (*La oligarquía) primera parte* (Mendoza: Best Hermanos, 1942); Pedro Santos Martinez, (editor) *Historia de Mendoza* (Buenos Aires: Plus Ultra, 1979); José Luis Masini Calderon, *Mendoza hace de cien años* (Buenos Aires: Ediciones Teoría, 1967); Agustin Alvarez, *Breve historia de la provincia de Mendoza* (Buenos Aires: Talleres de Publicaciones, 1910); Natalio Bontana, *El orden conservador; la política argentina entre 1880 y 1916* (Buenos Aires: Sudamericana, 1916).
3. *La Carcajada*, 11 April 1886.
4. Pedro Santos Martínez, "Mendoza, 1862–1892. Ensayo de interpretación sociopolitico," in *Contribuciones para la historia de Mendoza*, ed. Pedro Santos Martínez (Mendoza: Universidad Nacional de Cuyo, 1969): 131–173; Scobie, *Argentina*, 112–115; Rock, *Argentina*, 79–131.

5. For more on the economic plans of the Generation of 1880 see: Bontana, *El orden conservador*; Thomas F. McGann, "The Generation of Eighty," *The Americas* X: 2 (October 1953): 141–157; Oscar Cornblit, Ezequiel Gallo (h.), and Alfredo O'Connell, "La generación del 80 y su proyecto: antecedentes y consecuencias," *Desarrollo Económico* I: 1(enero-febrero 1962) : 641–691; Paul B. Goodwin, Jr., "The Central Argentine Railroad the Economic Development of Argentina, 1854–1881," *Hispanic American Historical Review* 57:4 (November 1977): 613–632.
6. William Fleming, "Regional Development and Transportation in Argentina: Mendoza and the Gran Oeste Railroad, 1885–1914," Ph.D. dissertation (Indiana University, 1976); Joan E. Supplee, "Provincial Elites and the Economic Transformation of Mendoza, Argentina, 1880–1914," Ph.D. dissertation (University of Texas, 1988); George Heaps-Nelson, "Argentine Provincial Politics in an Era of Expanding Electoral Participation: Buenos Aires and Mendoza, 1906–1918," Ph.D. dissertation (University of Florida, 1975).
7. Pedro Santos Martínez, "Vísperas y Repercusiones del 90 en Mendoza," in reprint of *Boletin de la Academia Nacional de la Historia* XLI (Buenos Aires, 1968): 15–22; Funes, Gobernadores de Mendoza, primera parte, 354–361. República Argentina, Archivo General de la Nación Colleccion Miguel Juárez Celman, Legajos, 161–164.
8. Santos Martínez, "Vísperas," 34.
9. Santos Martínez, "Vísperas," 22–38.
10. Supplee, "Provincial Elites," 8–11.
11. *Los Andes*, 2 January 1895.
12. Lucio Funes, Gobernadores de Mendoza (La oligarqia), segunda parte (Mendoza 1951): 36; Ramona del Valle Herrera, "Desde Caseros hasta el fines del siglo XIX," in *Historia de Mendoza*, Santos Martínez ed. 152–153; *Los Andes, cien años de la vida mendocina* (Mendoza, 1983): 28–29.
13. Supplee, "Provincial Elites," chapter 5.
14. Information concerning wages can be found in *Los Andes* during the period and in Mendoza Province, *Memoria descriptiva y estadística de la provincia de Mendoza* (Mendoza, 1893): 97–105; for more information on Lencinas's connection with the working class see Dardo Olguín, *Dos políticos, dos políticas* (Mendoza, 1956); Celso Rodrígues, *Lencinas y Cantoni* (Buenos Aires: Editorial Belgrano, 1979); Dardo Olguíon, "Los Lencinas: los gauchos de Mendoza," in *Los caudillos de este siglo*, ed. Félix Luna (Buenos Aires: Todo es Historia, 1976), 1–54; Pablo Lacoste, *La Unión Cívica Radical en Mendoza y en la Argentina 1890–1946* (Mendoza: Ediciones Culturales de Mendoza, 1994).
15. Once the election was certified, the opposition candidates relied on their standing within the landowning elite to ensure that legislators would seat them. Unfamiliar with such defiance on the part of the Mendozan electorate and wary of the political consequences of refusing to seat these two influential men, the UP relented and allowed them to take their seats. Mendoza Province, Registro

oficial de la provincia de Mendoza del año 1902, vol I [hereafter cited as ROM and year] (Mendoza: 1903), 145.

16. Paul Yves Denis, "San Rafael: La ciudad y su region," *Boletín de estudios geográficos* 16 (1969): 263.
17. *La Prensa*, 20, 22, 25 March, 11 June 1904; *Los Andes*, 11 August 1904.
18. *La Prensa*, 13 October 1903; 21, 22 July; 27, 29 August 1904.
19. Ibid.
20. Mendoza Province, ROM 1904 I, 355; *La Prensa*, 12 December 1904, 28 April, 25 November 1905.
21. Archivo Histórico y Adminstrativo de la Provincia de Mendoza [hereafter cited as AHM], Sección Independiente, Carpeta 2 de la biblioteca, "Leyes de Presupuestos y Gastos;" Mendoza Province, ROM 1904–1906; *La Prensa*, 19 November 1905; 4 March 1906.
22. *El Comercio*, 6 February 1904; *La Prensa*, 23 July 1900, 9 March 1904.
23. *La Prensa*, 10 October 1906.
24. *La Prensa*, 7, 8, 20, 22 August; 1, 5, 22 September, 8 October 1906.
25. Supplee, "Provincial Elites," 126–127; La Prensa, 22, 28, 29 October, 27 December 1906; Donald Peck, "Argentine Politics and the Province of Mendoza, 1890–1916," Ph.D. dissertation (Oxford, 1977): 86–87; Funes, *Gobernadores de Mendoza*, segunda parte, 110.
26. For more detail on the relationship between Civit and Figueroa Alcorta see: Donald Peck, "Argentine Politics and the Province of Mendoza, 1890–1916," Ph.D. dissertation (Oxford, 1977): 95–97, 135–138.
27. La Prensa, 30 October; 5, 8 November, 28 December 1907; Funes, Gobernadores de Mendoza, segunda parte, 114; Ana María Mateu de Pedrini, "Emilio Civit y el progreso de Mendoza," *Revista de la junta de estudios históricos* Epoca II, 10 (1984): 223.
28. Supplee, "Provincial Elites," 126–127; *La Prensa*, 22 October 1906; 29 October 1906; 27 December 1906; Donald Peck, "Argentine Politics and the Province of Mendoza, 1890–1906," Ph.D. dissertation (Oxford, 1977): 86–87; Funes, *Gobernadores de Mendoza*, segunda parte, 110.
29. *Los Andes*, cien años, 51; Funes, *Gobernadores de Mendoza*, segunda parte, 115, 118–119; Heaps-Nelson, "Argentine Provincial Politics," 141–144; *La Prensa*, 10 April 1908; 26 March 1909.
30. George Heaps-Nelson thought that the number of men under arms was far higher. See Heaps-Nelson, "Emilio Civit and the Politics of Mendoza," in John F. Bratzel and Daniel M. Masterson, *The Underside of Latin American History* (East Lansing, Michigan: 1977), 14; Funes, Gobernadores de Mendoza, egunda parte, 110; *Los Andes*, 12 October 1906, 21, 23, 24, 27 February 1906.
31. Funes, Gobernadores de Mendoza, segunda parte, 117; George Heaps-Nelson, "Emilio Civit," 15; ROM 1907, I; 468; *La Prensa*, 29, 30 August; 11 September 1907.

32. *La Prensa*, 28 September, 5 December, 9 December, 10 December, 14 December, 1910; Peck, "Argentinian Politics," 149.
33. *La Prensa*, 12 December, 14 December, 24 December, 25 December, 1910; 4 January 1911; Peck, "Argentinian Politics," 149. Peck claims that Ortega quickly rescinded the slaughterhouse contract, but he did not move against it until March 1912, after he had built his own political support.
34. Water concessions from Mendoza Province, ROM 1910; Los Andes, Cien Años, 54, 57; Decreto reglamentando la ley número cuatrocientos treinta en lo referente a concesiones de aprovechamiento de desagües y sobrantes de canales, Carpeta 68 de la biblioteca, Legislación de Aguas, AHM; Chaca, Breve historia, 41; Mendoza Province, ROM 1910, II, 525.
35. Although the reports in Los Andes and La Prensa give the impression that the water was being channeled out of the Diamante River into the Atuel, Peck states that the plan was to divert water from the Atuel to the Diamante. A map of the plan in *La Prensa* (6 May 1911) appears to support Peck's conclusion. No matter which way the water ran, important landowners downstream would suffer. See Peck, "Argentinian Politics," 150; Mendoza Province, Legislación fundamental; 15065; *La Prensa*, 17 February, 18 February, 2 March, 22 April; 6 May, 19 May, 1911; *Los Andes*, 22 March, 21 April, 1911:1.
36. *La Prensa*, 22 April 1911; Decreto 27, setiembre 1911; Decreto 9, noviembre 1911 noted as documents 50 and 51 in Marta Páramo de Isleño, Magdalena Alonso de Crocco, and Adolfo Cueto, "Aporte documental para una historia de irrigación del sur de mendocino," *Revista de historia de América y Argentina* X (1978–1980): 81; Peck, "Argentinian Politics," 149; Mendoza Province, ROM 1910, II: 50; Argentine Republic, Ministerio de Agricultura de la Nación, Junta Reguladora de Vinos, Recompilación de leyes, decretos y disposiciones sobre la industria vitivíncola (Buenos Aires, 1941): 459–463.
37. *Los Andes*, 5 May, 18 May 1911; Funes, Gobernadores de Mendoza, segunda parte, 142; Peck, "Argentinian Politics," 149.
38. Supplee, "Provincial Elites," 391–395; Peck, "Argentinian Politics," 151–156.
39. *La Prensa*, 9 November, 13 November, 17 November, 22 November, 1912; 18 January 1913.
40. *La Prensa*, 3 April, 11 April, 19 April, 16 July, 19 July, 23 July, 1913; 24 June, 1914; *Los Andes*, 23 August, 1913; *Il Tricolore*, 26 August 1913; 29 September 1913; Heaps-Nelson, "Argentine Provincial Politics," 84–85; Peck, "Argentinian Politics," 236–237.
41. *La Prensa*, 9 November, 13 November, 17 November, 22 November 1912; 18 January 1913.
42. For more information on the Lencinas phenomenon in Mendoza, see: Dardo Olguin, *Dos políticos, dos partidos. Emilio Civit y José Nestor Lencinas* (Mendoza: 1956); Celso Rodríguez, *Lencinas y Cantoni: El populismo cuyano en tiempos de Yrigoyen* (Buenos Aires: Editorial Belgrano, 1979).

CHAPTER THREE

Hipólito Yrigoyen, Salta, and the 1928 Presidential Campaign

Nicholas Biddle

Having won a 2-to-1, landslide victory, President Hipólito Yrigoyen looked over the cheering crowd at his inauguration on October 12, 1928. Turning to a friend he reportedly said, "These same people, in less than three years, will throw me out of power."[1] Less than two years later, Yrigoyen's prediction came true on September 6, 1930, when street crowds cheered General José F. Uriburu as he marched to the Presidential Palace to overthrow the government. This anecdote, related by Yrigoyen's biographer, Manuel Gálvez, signals the Argentine statesman's astute apprehension of the volatile political conditions generated in the 1928 presidential campaign.

This chapter seeks to explain Yrigoyen's titanic victory. Existing accounts remain superficial and misleading for two reasons. First, historians focus on Buenos Aires as the primary theatre of the campaign.[2] However, the 1928 campaign was unprecedented in that it took shape in the interior province of Salta and moved eastward to the great metropolis in a series of de facto presidential primaries. Secondly, the impact of events surrounding the September 1930 coup d'etat have overwhelmed historiographic concern with the genesis of the 1928 election. A sharp contrast has developed in the literature between Argentina's "experiment with democracy" (1912–1930) and the "infamous decade" (1930–1943) that followed.[3] The former is characterized as a muddled, benign, and generally progressive era, while the latter is openly coercive and fraudulent. In this context, the 1928 election was a closing moment of democratic success before the downfall of 1930.[4] What follows below argues that the electoral campaign exposed to the nation a pattern of fraud and coercion presumed by many to have passed into history. Violence mounted, and voters responded by sending Yrigoyen back into office. Simultaneously, however, reactionary forces initiated efforts that ultimately culminated in the presidential overthrow of 1930.

The most explosive issue identified with the campaign was oil nationalism. In September 1927, the lower house of Congress passed a bill to nationalize oil fields leased by the Standard Oil Company of New Jersey in the provinces of Salta and neighboring Jujuy.[5] It marked an unprecedented effort to expropriate the property of a foreign firm.[6] Never before had Yrigoyen or any member of his political party proposed confiscating private property, but the question of oil caused them to make an exception. The oil plan incited great wrath among elites. One conservative senator denounced it as "at base ... a war against the social structure."[7] Indeed, the plan was controversial, yet historians have neglected the origins of such controversy. They have done so because the events precipitating the Yrigoyenista stance took place in the tropical forests at the northwestern extreme of the nation.[8] Historical research in that distant region is scant and sources are limited.[9] Drawing extensively from provincial newspapers, this chapter describes the episode that catalyzed oil nationalism and subsequently set the stage for the 1928 campaign. Popular support for expropriating the Standard Oil Company marked a strong anti-imperialist position unusual in a nation noted for its open markets and foreign trade. The peculiar nature of events precipitating oil nationalism helps explain the force of this atypical political movement.

By itself, however, oil nationalism did not carry Yrigoyen to the presidency as is implied by David Rock, Carl Solberg, and others. Rather, a mounting wave of electoral violence presaging the politics of the 1930s determined that outcome. Oil nationalism initiated the campaign and drew the Yrigoyenistas (members of the *Unión Cívica Radical*, Radical Civic Union, still loyal to Hipólito Yrigoyen's leadership) into Salta's provincial politics. Moving a presidential campaign into the interior was an unusual step. What Yrigoyen's campaign managers encountered once involved in that relative backland was a revanchist conservative regime unwilling to tolerate the basic tenets of the celebrated 1912 Sáenz Peña Law stipulating universal male suffrage and secret balloting. Adjusting their strategy adroitly, Yrigoyen's lieutenants were able to champion the moralistic principles of suffrage and fulfillment of the constitution that had driven the Radical Civic Union since its inception.[10] Salta's conservatives responded with brutality. Eventually, Yrigoyen's followers answered in kind. By the time of the election, all civility had vanished and over 200 people had died in electoral violence.[11] Voters, however, understood that Yrigoyen stood for greater political freedom and civil rights. In effect, the mandate they gave him was a plea for authentic democracy.

The comfortable dichotomy between a democratic period opening with the Sáenz Peña Law succeeded in 1930 by an era of coercive, fraudulent government obscures a political continuity experienced in the Argentine interior. Though between 1912 and 1930 "the worst kinds of political skullduggery were banished,"[12] in Buenos Aires and the surrounding Littoral, as David Rock puts it, the same was not true in other regions. Conservative resistance to free and fair suffrage never faltered where the national press was not looking or executive intervention was not pending. This reactionary condition held sway in large sectors of the nation throughout the Sáenz Peña years. Historians have identified the pirate regimes of the Lencinas family in Mendoza and the Cantonis in San Juan alternately as divergent populist alliances or brutal throwbacks to nineteenth-century *caudillismo*.[13] However, these cases are painted as exceptions to a process of democratization advancing elsewhere. But this assessment overlooks politics in Córdoba, Tucumán, Jujuy, Salta, and other interior provinces. Because singular circumstances in 1927 compelled the *yrigoyenistas* to campaign directly in these locales, citizens across the nation read in the press of the recalcitrant nature of politics outside the Littoral. Popular reaction to political elitism and repression spread and became the key factor influencing the 1928 election. Fraud, violence, and a determined subversion of democracy characterized partisan competition in 1928, just as was the case throughout the nation in the 1930s. Contrary to the historiography, political conditions associated with the 1930s were already prevalent in 1928.

Circumstances of the 1928 Campaign

The 1928 election was exceptional. Hipólito Yrigoyen, 74 years old at the start of the campaign, hardly campaigned. Rumors of his senility circulated widely in the months preceding the April 1, 1928 vote, as did reports of his inclination to drop out of the race.[14] Having since 1896 led Argentina's most popular political party, the Radical Civic Union (*Unión Cívica Radical* or *UCR*), Yrigoyen first captured the presidency in 1916. The substance and style of that administration (1916–1922), however, split the party.[15] By 1924 those party members who remained loyal to Yrigoyen came to be called Personalist Radicals (or Personalists) and those who did not became Antipersonalist Radicals (or Antipersonalists).[16] Antipersonalists joined with conservatives in April 1927 to present a single presidential party ticket against Hipólito Yrigoyen. This alliance formed when participants at a Convention of the Parties of the Right agreed not to nominate their own candidates but, rather, to support the Antipersonalist ticket of Leopoldo Melo and Vicente Gallo.[17]

The Conservative–Antipersonalist coalition, called the United Front (*Frente Único*), believed itself "certain of triumph" at its inception.[18] Through the following year leading to the election on April 1928, the United Front compiled "[l]ong lists of supporters from business and banking, publicity tours, publications, meetings, and all kinds of promotional resources in vast scale made available with truly extravagant technical and financial means, [that] constituted an exhibition of power never seen before."[19] Meanwhile, as Felix Luna describes, the Personalists "hardly made a speech."[20] Moreover, the Personalist-Radical Civic Union delayed convening a nominating convention until just ten days before the election itself.[21] The venerable Yrigoyen chose not to attend the convention and, instead, sent a letter accepting the presidential nomination. In essence, then, without formally campaigning at all against a bipartisan coalition candidate, Yrigoyen swept the election by a 2-to-1 margin of victory in the heaviest voter turnout to that point in Argentine history.[22]

Yrigoyen's lack of official status as a candidate until ten days before the balloting did not shake the widely-held assumption that the old war-horse of Radicalism would make one last run. His formal noncandidacy enabled him to organize party affairs from an office in his home. Notoriously reclusive (detractors nicknamed him *el peludo*, or "the ant-eater"), Yrigoyen spent most of his political career in the confines of an office where his singular charisma and powers of persuasion on individuals or small groups had the most impact. His memory for people was prodigious, and he seemed never to forget a name or the circumstances of an individual's life. Having reached middle age before 1900, Yrigoyen almost never used the telephone or listened to the radio. He was an eccentric politician on the cusp of a new age with habits that ran contrary to the times.[23] Yrigoyen's idiosyncrasies matched the irregularity of Argentine elections. National, provincial, and municipal elections took place haphazardly, almost at random. Dates were not synchronized for balloting at different levels of office. Instead, each jurisdictional body fixed its cycle for electoral renewal subject to factional disputes, boycotts, and executive interventions.[24] For example, federal interventions in the province of Salta in 1918 and again in 1921 disrupted the pattern of gubernatorial succession. Governors were elected in December 1918, January 1922, November 1924, and December 1927. Presidential elections occurred in April 1922 and April 1928. Meanwhile, municipal elections in Salta's provincial capital occurred bi-annually in May only weeks after the equally frequent selection of national deputies. At times electoral campaigns seemed continuous. Oddly, the lack of definition to campaign seasons complemented Yrigoyen's style of constant home-based organizing. In a sense, he was always campaigning in a nation that itself was almost always involved in a campaign.

These conditions make it difficult, therefore, to mark the start of the 1928 presidential campaign. Carl Solberg suggests that the "petroleum debates of 1927 coincided with the opening of the presidential election campaigns for 1928."[25] His observation refers to Congressional deliberations in July and August 1927 over legislative proposals to nationalize the country's burgeoning oil industry. Though oil was a topic of frequent legislative consideration, the 1927 oil debate proved particularly volatile. In order to place the debate in an appropriate context, it is necessary at this juncture to diverge briefly into the history of oil production in Argentina.

In 1907 engineers in search of water discovered oil in the southern outpost of Comodoro Rivadavia, Chubut territory. Almost immediately politicians argued the pros and cons of nationalizing oil production. Anti-imperialism fueled the rhetoric. "It seems clearly proven that the revolution that has placed Mexico on the edge of abyss," argued one proponent in 1913, "has been nothing other than one of the many business deals transacted by the Standard Oil Company."[26] Fear of foreign domination underlay nationalization efforts for decades. Compromise legislation in 1910 established large government reserves in the Comodoro Rivadavia region and a state enterprise, *Yacimientos Petrolíferos Fiscales* (*YPF*), to develop them.[27] Private firms competed in uncontrolled zones, most prominently the Royal Dutch Shell Company. Reviled by Argentine politicians as an instigator of intrigue and corruption in Mexico, the Standard Oil Company of New Jersey made a weak foray into Comodoro Rivadavia, but to no avail.[28] Then, in the early 1920s, Standard Oil gained concessions from the Bolivian government to explore over one million hectares along that country's southern border. Geologists established that reserves stretched into neighboring Argentina. Quietly, the company began acquiring development rights in the northwestern provinces of Salta and Jujuy. By 1925, Standard Oil held leases for 143,000 hectares there.[29]

Legislators launched attempts to nationalize oil reserves in 1908, 1910, 1917, and each year after 1923 until the controversial debates of 1927.

President Marcelo T. de Alvear (1922–1928), who otherwise refused to intervene in provincial affairs, considered oil too vital to leave under the control of provincial authorities.[30] Alvear sent an annual message to Congress expressing his intention to "attribute to the nation possession of all mines in all the land, replacing the provinces in this management and authorizing in a permanent manner the power to exploit these mines when it [the federal government] deems advantageous."[31] Every year, congressmen more worried about alienating foreign capital than about falling victim to it, held sway to stalemate nationalization initiatives. Frustrated but unswerving, oil nationalists lurked in search of new weapons to overcome the opposition.[32]

In 1926, Standard Oil and members of the conservative party in Salta, called the Provincial Union, provided oil nationalists with the kind of scandal they were looking for to elevate their cause.[33] In April, three Standard Oil employees driving to northern Salta were ambushed on the highway, robbed, and killed. The North American oil firm responded as if it was an Argentine law enforcement agency. Wanted posters promising a 5,000-peso reward for the capture "dead or alive" of the murderers circulated the province.[34] The conservative Salta governor, Joaquín Corbalán, silently assented in the company's action. Predictably, hundreds of deputized vigilantes joined rural police in a massive, unrestricted manhunt. Days later a trigger-happy posse shot down an innocent scientist from France who had been studying the flora and fauna of northern Argentina.[35] Soon thereafter, six other suspects were rounded up, four of whom were dangled from trees by ropes tied around their torsos in a makeshift plaza of an oil-workers' camp called Senillosa. For three days Standard Oil employees and local law-enforcement agents flogged the alleged culprits at will. One man died on the second day. Two were taken away unconscious and presumed dead. Authorities claimed that two others "confessed" to the crime. Publicly the case was closed although, in fact, the five surviving victims were never tried, and later they were released from jail.[36]

For more than a year, the isolation of the northwestern province, a Provincial Union cover-up, and considerations of political timing kept news of the Senillosa tortures from national exposure.[37] In June 1927, General Alonso Baldrich, a former administrator in the state oil enterprise, *YPF*, and a close friend of its director, General Enrique Mosconi, broadcast the episode in a public conference entitled, "The Rivalry of Imperial Powers over Oil and its Consequences for Argentine Sovereignty."[38] Baldrich hammered home one theme in his address: "danger comes from the United States."[39] He drew a lesson from the events in Salta, that by "discarding the serene processes of justice. … Standard Oil exercise[d] a veritable dictatorship with its unnerving and corrupt influence." Baldrich's speech was electric. Righteous indignation resounded. Standard Oil was scandalized. For weeks his oration echoed on radio and in theaters across the nation. The newspaper editor in Salta whose daily first broke the story wrote ecstatically that, "All Argentina is on its feet in open battle against the Standard Oil Company."[40]

As intended, General Baldrich catalyzed immense popular momentum. Support for oil nationalism surged to its highest level just as the chamber of deputies opened debate on July 15, 1927 over two bills outlining principles and procedures for nationalizing the petroleum industry.[41] Until that moment, oil nationalism had been the bailiwick of Antipersonalist Radicals. As appointees of President Alvear, Generals Baldrich and Mosconi had become its standard-bearers. But Yrigoyen saw political opportunity in the moment, and soon after July 15 he called together the Personalist bloc of deputies (59 of 158 seats) to direct them to steal the thunder from the Antipersonalists. His instructions were simple: become fiercer oil nationalists than the Antipersonalists. In the discourse of the debate, those instructions translated into championing the wholesale expropriation of private oil operations.[42]

The new Personalist position was sufficiently extreme to distinguish the party from Antipersonalists who favored mixed enterprises without expropriation. But the Personalist tactic also incurred the wrath of the private sector and conservative politicians. Never before had Yrigoyen or his party suggested tampering with private property. Opponents considered the proposal a complete departure from ruling assumptions of the liberal economy and the standard position of the Radical Party. Congressional debate became "agitated and furious."[43] Undaunted, one Personalist, Deputy Alcides Greca, openly embraced "scientific socialism":

> The monopoly which we propose, this monopoly of the new industrial state, arises by virtue of that economic, political, and philosophic tendency that places into the hands of the State the great industries and the large public services, to temper, in a certain manner, the effects of capitalism and the struggle of the classes. The celebrated manifesto of the two founders of scientific socialism says that: "the proletariat will use its political supremacy to wrest, by degrees, all capital from the bourgeoisie, to centralize all instruments of production in the hands of the State, i.e. of the proletariat organized as the ruling class; and to increase the total of productive forces as rapidly as possible." This resembles somewhat that which we are doing: putting in the hands of the State one of the factors of production.[44]

International developments strengthened the Personalists' hand. News of the United States invasion of Nicaragua in late July 1927 drew an acrimonious response from all sectors of the public and Congress. Simultaneously, working class protests against the United States accelerated as the date for executing the anarchists Sacco and Vanzetti approached. Meanwhile, a recent ban on Argentine meat imports by United States officials, ordered to prevent the spread of hoof-and-mouth disease, rankled upper-class cattlemen.[45] Widespread anti-United States sentiment compelled retaliation; oil legislation became the weapon of choice. Recognizing the benefits of compromise, Personalists stepped back from their demand for expropriation to join with Antipersonalists in passing a measure that placed all subsoil minerals under federal jurisdiction, created a state monopoly on petroleum transport, and prohibited oil exports. The 88-to-17 vote was overwhelming. Emboldened, Personalists renewed their militancy by introducing an amendment to give the state oil enterprise, *YPF*, a monopoly on all rights to future exploration and production. Passing by a far slimmer margin of 65 to 55, the Personalist amendment pushed the limit of acceptable state intervention for deputies in the chamber.[46] Given a more elitist makeup in the Senate, everyone recognized that the militant Personalist plan stood no chance of reaching President Alvear's desk for signing into law during the current legislative year.[47] This political fact caused Yrigoyen's detractors to allege that the whole initiative reflected pure opportunism.[48] While the evidence corroborated that observation, nevertheless, it was effective politics.

The oil issue became "a great bandwagon movement in Yrigoyen's favour."[49] Historians find it the single substantive issue of the 1928 presidential campaign, which "the [P]ersonalist machine skillfully used" to deliver "one of the most sweeping landslides in Argentine history." But they provide little detail. Seven months separated the end of congressional oil debate on September 8, 1927 from presidential balloting on April 1, 1928. Who championed oil nationalism and who contested it in that interim? In what forums was it debated? How did the "bandwagon movement" gain momentum through seven months?

The Campaign in Salta

The United Front coalition outspent the Radical Party enormously where generally it counted most, in the densely populated metropole of Buenos Aires and the surrounding provinces.[50] The one setting that the Personalists identified as propitious for carrying on the oil debate was in Salta, where the scandal that launched the Personalist platform had taken place. Reflecting the haphazard nature of the Argentine electoral process, Salta was also the first of four provinces scheduled to hold gubernatorial elections between December 1927 and the presidential balloting in April 1928.[51] Yrigoyen and his advisors saw the opportunity to transform these gubernatorial races into presidential primaries of sorts. Choosing to avoid taking on the United Front head-to-head in Buenos Aires, Yrigoyen announced to a reporter in early September that in Salta he would "risk the fate of the upcoming electoral campaign for the presidency."[52] Yrigoyen's declared focus on the interior province marked a significant departure from previous Radical campaigns in which a "tight control over party committees" in and around Buenos Aires enabled a complex distribution of patronage to maintain allegiance from the urban majority.[53] But Personalist extremism on the oil issue compelled the party to adopt unusual tactics. There were great risks involved in propounding the expropriation of corporate property. In Salta, Yrigoyen hoped to keep the public aware that the logic of oil nationalism sprang from violent excesses perpetrated by Standard Oil, with the collusion of the province's conservative government.

The first manifestation of Yrigoyen's strategy came in the form of a delegation of trusted advisors dispatched to Salta in late September to prepare the gubernatorial campaign.[54] Deputy Diego Luis Molinari headed the group, comprised mostly of members of the Personalist National Committee. Molinari's presence added prestige to the contingent, since most Radicals considered him the heir-apparent to Yrigoyen.[55] Gleefully, Salta's partisan Personalist newspaper, *La Voz del Norte*, reported that the delegation's presence "absolutely invalidated all that has been said about the indifference with which Buenos Aires has looked upon the political affairs of this province."[56] However, not all local Personalists were pleased with the delegation's arrival.

Already Yrigoyen had meddled in the process of selecting a gubernatorial candidate for the province, exhibiting the tendency to autocratic control over party affairs that his enemies reviled. Without consulting Salta's Personalist leaders nor waiting for a nominating convention, Yrigoyen had begun in June 1927 to pressure a former Radical governor of Salta, Adolfo Güemes, to throw his hat into the ring. While governor from 1922 to 1925, Güemes had alienated provincial Radicals by conforming too closely to conservative party interests and, therefore, he was hesitant to rejoin the fray.[57] His attractiveness was that, during his tenure, Güemes openly opposed and effectively retarded Standard Oil's expansion in the province.[58] In the context of the current campaign, that position alone was sufficient to qualify him in Yrigoyen's eyes. Disregarding local objections that a Güemes candidacy was procedurally irregular and political suicide, Yrigoyen repeatedly tried to coax the former governor out of retirement.[59] Meanwhile, Molinari and his assistants stalled further progress in the provincial nominating process. Eventually in late October, the vacillating Güemes unconditionally refused the candidacy.[60]

With less than five weeks remaining before the gubernatorial balloting on December 4, Deputy Molinari convened Salta's Personalists to emergency meetings on November 2. Proclaiming full control over "the general direction of the Party throughout the province," Molinari rejected holding a nominating convention in the interest of time. After three days of talks Molinari emerged on November 4 to announce Julio Cornejo the Radical Party nominee. Acting to assuage ruffled sensibilities among provincial Radicals angered at their loss of local control, Molinari praised the "perfect cohesion" of Salta's Personalists and emphasized how important "the current electoral labors" were because of the circumstances that made it "a prologue to the presidential election of 1928."[61] Yet, a sense of party disarray lingered.

Julio Cornejo had recently converted to Radicalism in the wake of the Standard Oil scandal. Because he had been a Provincial Union member for two decades before then, many provincial Radicals questioned Cornejo's motives and the depth of his conviction.[62] These concerns were swept aside under Molinari's command. Like Adolfo Güemes, Cornejo's attraction to the Personalist National Committee was his oil nationalism. Cornejo had proven his mettle in the provincial legislature since 1926, when he condemned Jersey Standard and its conservative allies repeatedly.[63] Nevertheless, Cornejo now turned his attention to the short campaign ahead with only grudging support from local party officials.

The Provincial Union was buoyed by the circumstances and confident that "with the elimination of the Güemes candidacy they would pass into first place in the race."[64] Too little time remained for debate over the oil issue to sway local opinion on either side. Since 1922, moreover, conservatives in Salta had renewed mechanisms of electoral coercion that predated the rise of Radicalism. These involved compelling rural workers to vote conservative or risk unemployment and homelessness.[65] As Personalists scrambled to identify a nominee, conservatives anticipated that the rural votes they could coerce would suffice to carry them to victory as, indeed, had been the case in the previous gubernatorial election of 1924.[66]

The unpredictable factor in the Salta campaign was the role of Antipersonalists. Though linked to conservatives nationally through joint support for the Leopoldo Melo-Vicente Gallo presidential ticket, Antipersonalists in Salta refused to ally with conservatives at the provincial level. Neither did they feel strong enough to launch their own gubernatorial candidate. After considerable hedging, party leaders decided to reunite tacitly with the Personalists by supporting the Güemes candidacy. They did so, however, just eight days before Dr. Güemes renounced his candidacy.[67]

That unfortunate timing did not prevent the first step toward Antipersonalist-Personalist reconciliation at the provincial level on the basis of equal enmity for the Provincial Union's gubernatorial candidate, Manuel Alvarado.[68] Nine years earlier, Alvarado had run for governor on a thinly veiled program to undermine universal male suffrage.[69] Now, Antipersonalists denounced his candidacy as an "immoral insult."[70] Since both Antipersonalists and Personalists considered suffrage and fair elections the essence of Radicalism, the unintended consequence of the Alvarado candidacy was to build a bridge between the divided factions of Radicalism in Salta.

Yrigoyen's lieutenants quickly realized the implications. Though oil nationalism galvanized the Personalist Party through the 1927 oil debates, its immediacy faded from the electorate's attention as the bills passed in the lower house sat in limbo during congressional recess. The Alvarado candidacy, on the other hand, assaulted democratic sensibilities cultivated by the Radical Civic Union for a generation. Already in November 1927, conservative strong-arm tactics pervaded the Salta campaign. Deputy Molinari and his two closest associates, Federico Beschtedt and Carlos Borzani, hoped that the same disgust at conservative skullduggery that effectively rejoined Antipersonalists and Personalists in Salta might have a similar impact nationally. Their task, then, was to make conservative *revanchismo* in Salta a national issue.

The Personalist envoys adopted a media strategy. Federico Beschtedt and Carlos Borzani began writing daily dispatches outlining electoral violations across the province. Common forms of intimidation they described included: public employees forced to attend conservative party functions; Radicals arrested and detained for hanging campaign posters; local Radical Party headquarters attacked and damaged; and routine threats and beatings of Radicals.[71] Personalists sent these dispatches to the Buenos Aires press, Personalist National Headquarters, and federal officials. The largest national newspaper, *La Prensa*, initially downplayed the reports as predictable byproducts of electoral competition, much as bruises and broken bones accompany the sport of rugby. When one of Yrigoyen's delegates landed in jail in the northern district of Orán, *La Prensa* affected nonchalance, observing that the victims of electoral repression always "count among the opponents of the government," and tend to be pursued "with a rigor not given for thieves, assailants, or armed robbers," while inevitably as well, "police serve the political interests of incumbents."[72]

Salta's conservatives proceeded full steam ahead. Stumping for Alvarado, the Provincial Union's titular head, Robustiano Patrón Costas proclaimed the gubernatorial race a "patriotic crusade to impede the return of an ill-fated hour for the country."[73] United Front colleagues echoed his apocalyptic perspective elsewhere. Benjamin Villafañe, former Antipersonalist governor of neighboring Jujuy, and Matías Sánchez Sorondo, conservative senator and legal counselor for Jersey Standard, scorned Yrigoyen and the Personalist oil program in a conference held in the national capital during the November campaign. Villafañe accused Yrigoyen of demagoguery, embezzlement, and treason, predicting that his return to the presidency "would bring back the times of the *mazorca*," that is, the domestic terror exercised under the dictator Juan Manuel de Rosas. Meanwhile, Sánchez Sorondo spoke ominously about revolution and constitutional breakdown:

> The country does not need to defend itself against imaginary dangers, but it does need to defend itself against real ones, like these ill-advised, revolutionary, and anarchic laws. Yesterday it was rents, today it is oil, tomorrow it will be rural lands, threatened by redistribution. At base this is a war against the social structure and it begins by an attack against one of its strongest bulwarks: the right of property. These waves of oil, which according to the often cited phrase of Lord Curzon, carried the allies to victory, may drive us to constitutional ruin.[74]

Calumny leveled against Yrigoyen and the Personalist oil program was old hat by November 1927. While it reinforced the convictions of United Front partisans, such vilification also reflected an absence of new issues that might turn undecided voters against Yrigoyen. The aging Radical leader's aloofness and sedentary ways, in effect, worked to deprive the United Front campaign of content beyond emotional harangue and tired rhetoric. These conditions, however, did not diminish the vehemence of Yrigoyen's enemies, (as the subsequent coup d'etat in September 1930 indicates), nor the seriousness of their words.[75]

The Personalist strategy to turn conservative electoral corruption in Salta into a national issue began to take effect in the third week of November. Accounts reached the press of unarmed members of the Radical Party being shot and wounded by the sheriff and deputies of Orán, a town very close to the bulk of Jersey Standard's drilling operations. One victim of the assault described the local police as "reinforced with petroleum guards from the Standard Oil Company."[76] The danger of associating the North American firm with local police and political repression against Radicals induced *La Prensa* to chastise provincial conservatives in an article entitled "Poor Civic Education."[77] Beyond that, the newspaper called for federal intervention to pacify the province and guarantee fair elections.

The newspaper's appeal for federal intervention reflected genuine outrage in cosmopolitan Buenos Aires at electoral conditions in the interior. Elite acceptance of the Radical Party after 1912 depended upon the ability of representative democracy to legitimate the status quo, bind the middle class to the liberal economy, and bolster foreign investment confidence.[78] Electoral violence tainted the image and function of the political system, which had distinguished Argentina from its South American neighbors for 15 years. *La Prensa*'s aspersions reflected alarm at the gulf between democratic forms in Buenos Aires and their absence in Salta. Moreover, the Provincial Union's violence undoubtedly troubled Antipersonalist allies throughout the United Front. Seeking to reverse negative consequences, the Provincial Union issued a press release denying wrongdoing in general and shifting blame onto "reckless provocations effected in most instances by persons from outside the province who have come to act as electoral agents [i.e., Molinari and associates]."[79] But their rebuttals only raised suspicions and increased public attention. On the eve of the gubernatorial vote, *La Prensa* ascribed a "transcendence" to the campaign for having "awoken extraordinary expectation" across the nation.[80]

The unethical, coercive pattern of the campaign extended into election day on December 4. Voters woke up to find telegraph lines cut throughout the province. One press report in Buenos Aires attributed the damage to a "cyclone," but no account of bad weather surfaced in the province.[81] Personalists charged conservatives with the act "in order to hide the villainy" they planned.[82] In fact, Carlos Borzani got wind of sabotage beforehand and on December 3 alerted Interior Minister José Tamborini by telegram, but to no avail. The communication blackout prevented journalists from detailing the day's events until December 6, but the content of their reports was disturbingly familiar: Radicals jailed without charge, false votes circulated, and at one voting poll armed guards permitted only Provincial Union members to enter.[83] In the succeeding month that it took for officials to tabulate the votes, Radicals lost heart in the expectation that conservative fraud had been sufficient to steal the election.[84]

Federico Beschtedt sent a final dispatch to Personalist headquarters in Buenos Aires commenting that from the start "it was evident that they [conservatives] were preparing by reprehensible means a brutal show of force."[85] Meanwhile, Yrigoyen's mouthpiece, *La Epoca*, lamented the "profound subversion of all values occurring in the development of the nation's political contests at the present time."[86] The Personalist mood turned somber. Inversely, Salta's conservatives were so confident that they had Manuel Alvarado's portrait hung in the governor's office.[87]

Then "that accident happened in Salta,"[88] as one surprised observer put it. Julio Cornejo won by a slim margin.[89] Personalists in Buenos Aires were ecstatic. The Radical press heralded Salta's "humble workers" as having "fulfilled a high nationalist mission" that included "the defeat of Standard Oil."[90] Local party leaders telegraphed Yrigoyen to "proclaim this splendid triumph the first link in an uninterrupted chain that will carry you ... to the highest office of the Republic."[91] Having led his delegation from Salta to the next gubernatorial campaign in Tucumán, Deputy Molinari telegraphed Yrigoyen that news of the victory prompted Personalists and Antipersonists in that province to join forces. "It is a day of jubilation," he exulted, "which augurs another in January's [gubernatorial] contest."[92]

Three days after Cornejo's victory was known, *La Prensa* published an editorial decrying an absence of programmatic substance in the presidential campaign. Antipersonalists in the United Front appeared "contaminated with a virus of discord," and should "the spectacle of internal anarchy" continue, the daily speculated, United Front "failure in the upcoming elections" seemed certain.[93] Specific issues did not drive the political tide. Rather, the ground of contention became the comparatively recent and still pioneering democratic process launched by the Sáenz Peña Law. The gubernatorial election in Salta revealed how persistently provincial conservatives rejected free and fair elections. Undoubtedly, the "virus of discord" in the United Front stemmed from the contradiction between the commitment to free suffrage held by Antipersonalist Radicals and the direct or indirect opposition to it among conservatives.

In 1919, Provincial Union spokesmen openly attacked popular democracy using the rhetoric of social Darwinism:

> Argentine "oligarchies" ... follow the law of natural selection. This law accords to the superior man a natural right to exercise his authority over the intellectually amorphous mass who are not capable of thinking nor working for the good nor advantage of the community.[94]

By 1928, assaults on the democratic process turned bitter and incriminating. "The closed voting booth of which so much is said," declared one scornful conservative, "opens the way for many betrayals, ungratefulness, deceptions [and] disloyalties."[95] In essence, provincial conservatives defied democracy in favor of elitism and oligarchy through 16 years of the Sáenz Peña Law.[96]

In contrast, Radical Party pronouncements glorified "those who feel the truth of democracy and aspire for a government of the people and for the people."[97] As Gabriel del Mazo emphasizes, "[t]he Radical Civic Union was born as a reaction of public spirit against a total subversion of moral values."[98] Conservatives in Salta opened the way for Personalists to rally around that original standard, at least in terms of public relations. Yrigoyen, of course, was touted as the champion of the people and the free vote. Though partisans on all sides resorted to force as passions overwhelmed reason and specific issues faded from view, the allure of the concept of democracy tilted the balance of power in favor of Yrigoyen and the Personalist Party. As gubernatorial elections ensued in Tucumán and then Santa Fe and Córdoba, voters expressed their preference and, perhaps, their yearning for electoral gains formally achieved but still not firmly embedded in social practice. Yrigoyen's campaign gained decisive momentum at this juncture.

The Campaign after Salta

In late December Deputy Molinari moved his delegation to Tucumán to sustain the strategy forged in Salta. There Radicals began to respond to conservative coercion in kind.[99] When one partisan died at the hands of Tucumán police, a wave of denunciations and pleas for federal intervention ensued.[100] Exacerbating tensions, multiple "delegations" of conservatives and Antipersonalists, as well as Personalists, arrived in Tucumán from various provinces "to cooperate in the electoral campaign." Many members of these groups were not so much political operatives as provocateurs with disruption their primary objective. Commonly, "citizens of a different political tendency" arrived to heckle, cajole, and fight with the crowd during campaign speeches.[101] Press headlines declared, "Campaign Activities Turn Violent," and casualties mounted, often from gunfire.[102] At one rally two days before the Tucumán voting on January 15, 14 people were hospitalized after a brawl, while at another rally four others suffered bullet wounds.

As in the case of Salta, censures and petitions to President Alvear for increased federal protection fell on deaf ears. Interior Minister Tamborini issued a statement on January 13 that "at this opportunity the Executive reaffirms his intention to maintain absolute political neutrality" in the electoral affairs of the nation.[103] This sort of impartiality only encouraged further fury. In fact, President Alvear's waffling reflected a deep ambivalence. Hipólito Yrigoyen had been his mentor in the early days of the Radical Party and had selected him for the presidential candidacy in 1922. By that point, however, the Radical Party was splitting apart and Alvear took up the mantle of leadership in the dissident Antipersonalist faction. Undaunted by his underling's efforts at independence, Yrigoyen maintained a covert relationship with Alvear that prevented personal enmity between them from taking root. Nevertheless, when conservatives and Antipersonalists joined forces in the United Front, Alvear agreed in the "indisputable necessity of those who oppose the candidacy of Señor Yrigoyen to unite."[104] But as the campaign kicked off, it became clear that Alvear's support for the Melo-Gallo ticket was passive at best. By the time of Interior Minister Tamborini's January 13 press release, the official image presented of high-minded neutrality was understood by most people to mean presidential paralysis.[105]

Personalists gained a second victory in Tucumán and Molinari's entourage moved to Santa Fe for the third gubernatorial race. A few days before that vote, Santa Fe Governor Ricardo Aldao telegraphed a plea for federal intervention to stop the violence. "The whole province from one extreme to the other," he raged, "is scourged with malevolent people foreign to the province, of Personalist affiliation, many of whom, according to what has been able to be established, have been transported in vehicles with license plates from Buenos Aires." A number of assassinations had occurred and among the victims were two policemen. As an Antipersonalist, Governor Aldao laid full blame on the Personalists. The press was not so uniform in ascribing guilt, but was just as exasperated at the political situation. Politics were taking on the form of gang warfare:

> It seems that the sense of prudence and mutual respect have been lost and that the conviction has arisen that victory belongs to the party that carries out the most extreme assaults and unleashes the most inflamed passions.[106]

In the wake of Personalist victory in Santa Fe on February 5, a sense of futility filtered through the ranks of the United Front. The coalition's treasurer confided to a friend that he ought to return all campaign funds to contributors, so convinced was he of defeat.[107] Despondence at the prospect of electoral victory turned some to consider extra-legal strategies to stop Yrigoyen. Public rumor held that the Minister of War General Augustín Justo was plotting a preemptive coup d'etat.[108] General Justo felt compelled to issue a public denial that sounded more like a confession. "I believe," he wrote, "that we near a very difficult hour and I believe as well that men of state and all citizens ought to prepare themselves to avoid the ills that experience has taught us will come."[109]

By February's end, the United Front seemed to have lost all legitimacy. Worse, the political system itself was crumbling. Conspiracies and rumors of conspiracy complemented brutal displays of public violence on a daily basis.[110] The appearance of goon-squads, or groups of partisans determined to molest and assault political competitors, was perhaps the most alarming characteristic that emerged.[111] These were the precursors to the *Klan Radical* and the *Liga Republicana* who in the year ahead clashed frequently and bloodily, contributing to an "atmosphere of acute tension" that lasted into 1930.[112] In tones half-resigned and half-accusatory, *La Prensa* concluded "that the national government doesn't know how to guarantee the right of reunion," and that "only tolerance, weakness, or complacency explain the manner in which the [Alvear] government is proceeding before the breach of public order."[113] Simultaneously internecine squabbles mounted within the United Front for "continual and repeated errors that have permitted the figure of a mediocre caudillo [i.e., Yrigoyen] to be converted into an idol."[114] Sensing the irreversibility of the momentum swinging his way, on March 18, two full weeks before the presidential election, Yrigoyen called upon the party faithful to refrain from all campaigning in order to allay the situation before the actual voting.[115]

A few days later, the Personalist-Radical Civic Union opened its nominating convention. Its first act was to propose "as homage to the Radicalism of Salta (which had given in January the first signal of triumph) that the delegate of that province be selected president of the assembly."[116] This tribute underscored the success of the Personalist strategy hatched in Salta to turn conservative thuggery to Radical advantage one final time. Additionally, delegates from Tucumán and Córdoba were appointed assembly vice-presidents. When the time came to put forward nominations, only one name was heard: Yrigoyen.

Conclusion

Yrigoyen's sweeping victory is generally recounted as the result of his "magic sorcery," the subservience of party faithful, or as the outcome to popular support for oil nationalism.[117] Carl Solberg synthesizes these factors as follows:

> While Yrigoyen avoided the limelight, the Personalist machine skillfully used the oil policy debates of the previous year to present the masses with an image

> of a Radical Party that was leading the fight to prevent the rapacious Yankees from exploiting Argentina's petroleum resources.[118]

Solberg's assessment remains essentially uncontested. Nevertheless, this chapter argues that it glosses over the content and impact of the campaign. Ultimately, violence and false democracy became the determining factors of the election. At the start, oil nationalism mobilized public opinion and focused attention on Salta. Yrigoyen and his lieutenants shrewdly took advantage of the patchwork structure of Argentine elections to organize Personalists in Salta's gubernatorial race five months prior to the presidential vote. However, sending national figures from Buenos Aires to campaign in an interior province for a gubernatorial candidate was so unusual as to be oblique. Historians have overlooked the tactic completely. It signaled Yrigoyen's perception of disadvantage in the face of the United Front. He chose not take on his opposition directly in Buenos Aires but to work around them. As Etchepareborda points out, Personalists controlled only 2 of 14 provinces; moreover, governors in nine provinces declared support for the Melo-Gallo ticket before the 1927 oil debates.[119] The Personalist gambit in Salta contained an element of desperation.

Deputy Molinari's original mission was to sustain public scrutiny on the oil issue, but other circumstances overwhelmed that approach. What the Personalist officials found in Salta was an unreconstructed oligarchic regime maintained in the model of Argentine politics prior to the 1912 Sáenz Peña Law (and after 1930). Radicals were familiar with such situations, and the Molinari delegation quickly adjusted. Nevertheless, there was an eerie quality to the process, a sense of disbelief. At one point a member of the Molinari delegation appealed to the Provincial Union leader Robustiano Patrón Costas to "do nothing more than remotely approximate the electoral correction of the province of Buenos Aires."[120] His plea clearly went unheeded. As story after story of mounting violence reached Buenos Aires, alarm registered even in the conservative press. In exasperation *La Prensa* lamented that the "national government does not know how to guarantee the right of assembly."[121]

While historians recognize the incomplete institutionalization of the Sáenz Peña Law in the interior through the period of Argentina's so-called "democratic experiment," the dominant image projected by both contemporary media and subsequent historical literature is of an urbane and evolving political system.[122] Obviously, the Argentine business and political communities had much at stake in projecting to the international community an aura of political stability and progress. Confidence in the export economy, in large measure, depended upon it.[123] And arguably, within the Littoral region, politics were comparatively advanced in the context of Latin America at the time. However, this chapter maintains that the relative civility of political practice through the 1920s covered but did not replace a deeper tradition of fraud and coercion.

The single issue that elevated the intransigent Radical Civic Union into political legitimacy in 1912, free and fair suffrage, remained the bone of contention in 1928. Yrigoyen's imperious behavior, which had divided Radicals into Personalists and Antipersonalists and provided grist for conservative opposition, lost importance in the wake of the direct assault on basic political rights launched by the Provincial Union in Salta. That race unleashed a suppressed rage at the hypocrisy of Argentine democracy and caused partisans across the nation to vent their fury in the public sphere. The issue at hand was no longer oil nationalism, it was political legitimacy itself.[124] In province after province, politics turned disruptive: obstructing dialogue and provoking conflict. Each side viewed the other as fundamentally illegitimate. Ultimately, voters chose the party that promised them greater rights.[125] But their mandate emerged in a reactionary environment and many on Yrigoyen's side also besmirched those rights. Unfortunately, the roots of democratic habit proved too shallow to upend the unraveling, and a sense of political civility and progress did not return after the campaign.[126]

The nature of the 1928 campaign raises questions about the real content of Argentina's "experiment with democracy." Apparently there were two political spheres, the Littoral and the interior. In the former, elections took place in relative peace and legality, while in the latter much the opposite was true. In 1928, however, Argentines elected a president whose career was dedicated to guaranteeing full suffrage and free elections across the nation. The campaign revealed the Sáenz Peña Law had been a myth for large segments of the population. It also alerted elites in Salta, Buenos Aires, and elsewhere that the limited democratic system they were willing to tolerate was highly volatile when exposed to the light of day. The conspiracy to prevent Yrigoyen having success began not just in the 1928 campaign but because of it.

Yrigoyen's premonition of downfall—made at the height of his electoral triumph—brilliantly captured the fragile environment he faced looking over the inaugural crowd. Similarly portentous, soon before the inauguration, General José F. Uriburu confided to a Spanish American diplomat, "Yrigoyen will rise into the government, but he will not last because I will throw him out."[127] Unfortunately for the fate of Argentine democracy, both predictions came true.

Notes

1. While it is difficult to know the truth of this episode, its reporter Manuel Gálvez assures us of its veracity in the introductory essay of his excellent biography, *Vida de Hipólito Yrigoyen,* rev. ed. (Buenos Aires: Club de Lectores, 1975), 377, 13–25.
2. For instance, David Rock underscores that "the dependent middle class groups in the cities, which were the core of Yrigoyen's popular backing" carried the day. David Rock, *Politics in Argentina 1890–1930: The Rise and Fall of Radicalism* (London: Cambridge University Press, 1975), 238.
3. Argentina's "experiment with democracy" began with passage of the 1912 Sáenz Peña Law guaranteeing universal male suffrage and secret balloting. The "infamous decade" began with Yrigoyen's overthrow in 1930 and lasted to the 1943 coup d'etat that marked the rise of Juan Perón to national power. Peter Smith utilizes these phrases in *Argentina and the Failure of Democracy: Conflict among Political Elites, 1904–1955* (Madison, Wisconsin: University of Wisconsin Press, 1974), xv. The Sáenz Peña Law was the first of its kind in the continent and in step with democratic development in Europe. See Eric Hobsbawm, *The Age of Empire, 1875–1914* (New York: Vintage Books, 1989), 86–87. The stark division between an era of "democratic experiment" (1912–1930) and a subsequent "infamous decade" (1930–1943) has become commonplace. Perhaps the nationalist writers of the 1940s were most influential in establishing the dichotomy. See J.J. Hernández Arregui, *La formación de la conciencia nacional* (Buenos Aires: Plus Ultra, 1973), 290–293. Mark Falcoff and Ronald H. Dolkart articulate the dichotomy well in the preface to Falcoff and Dolkart, eds., *Prologue to Perón, Argentina in Depression and War, 1930–1943* (Berkeley: University of California Press, 1975), ix–xiv.
4. The fact of so overwhelming an electoral victory suggests an electoral system adjusted to the rules of representative democracy as, for instance, in the following rendering by David Rock, *Argentina, 1516–1982: From Spanish Colonization to the Falkland Wars* (Berkeley: University of California Press, 1985), 210. "In the 1928 election the Yrigoyenistas swept aside all opponents. Yrigoyen regained the presidency with around 60 percent of the popular vote, a victory that marked the acme of a public career which had spanned more than half a century."
5. Two bills were passed on September 1 and 8, 1927. The first established federal controls over all oil resources, gave the state a monopoly in transporting oil and prohibited oil exports. The second empowered the state oil enterprise, *Yacimientos Petrolíferos Fiscales* (*YPF*), to carry out all future oil exploration and production. The original form of this bill sought to expropriate existing private holdings but was amended to allow firms to maintain their productive operations. Argentina, Cámara de Diputados, *Diario de Sesiones*, vol. 4 (Sept. 1, 1927), 361–362; and (Sept. 8, 1927), 478.
6. David Rock notes that the oil plan isolated its attack on Standard Oil without turning its sights to Royal Dutch Shell or other firms. Also, Rock underscores the traditional Radical Party advocacy of economic liberalism in *Politics in Argentina,* 235–240.

7. This statement, uttered publicly in a November 1927 conference in Buenos Aires by conservative Senator Matías Sanchez Sorondo, was immediately published. Benjamin Villafañe and Matías Sánchez Sorondo, *La palabra de un patriota sobre el problema de la legislación del petróleo* (Buenos Aires: Imprenta Dominguez, 1927), 39.
8. As described below, vigilante violence executed by Standard Oil employees operating in Salta catalyzed the oil campaign. Carl Solberg devotes a paragraph to the incident in his book *Oil and Nationalism in Argentina* (Stanford: Stanford University Press, 1979), 120. Rock makes no mention of it specifically even though he discusses the genesis and impact of the 1927 oil debates in *Politics in Argentina*, 235–240. Gabriel del Mazo overlooks the events in Salta altogether in *El radicalismo* 3 vols. (Buenos Aires: Raigal, 1955), 1: 63–78. Carlos Mayo, O.R. Andino and F. Garcia Molino also fail to mention the episode in *La diplomacia del petroleo 1916-1930* (Buenos Aires: Centro Editor de América Latina, 1976). Enrique Mosconi, who was director of the state oil enterprise, *Yacimientos Petrolíferos Fiscales* (*YPF*) in 1926, reprints a portion of the 1927 speech by General Alonso Baldrich that exposed the vigilante torture, but does not integrate the information in his book *Y.P.F. contra la Standard Oil* (Buenos Aires: Castagnola e Hijo, 1958).
9. Contributing editor to the *Handbook of Latin American Studies*, Joseph T. Crescenti, observes progress in writing the histories of "up-river and interior provinces," but also observes that some provinces, "for natural and other reasons, lack the historical records for a completely satisfying account of their past." *Handbook of Latin American Studies*, vol. 50 (Austin: University of Texas Press, 1990), 330.
10. Curiously, David Rock perceives the oil program to have enabled *yrigoyenistas* to avoid normal appeals to "democracy" and "defence of the constitution." *Politics in Argentina*, 232.
11. Manuel Gálvez estimates that nearly 200 Radicals alone were killed without counting the numbers of conservatives and their allies who also fell victim to the violence. Gálvez, *Vida de Yrigoyen*, 370.
12. Rock, *Politics in Argentina*, 58.
13. See Celso Rodríguez, "Regionalism, Populism, and Federalism in Argentina, 1916-1930" (Ph.D diss., University of Massachussetts, 1974); and Dardo Olguín, *Lencinas, el caudillo radical, Historia y mito* (Buenos Aires: Vendimiador, 1961).
14. Yrigoyen's opponents suggested his senility routinely, however, a political ally, Deputy Diego Luis Molinari, inferred the same in Jujuy in January 1927. *El Intransigente* (Salta), 27 January 1927. This rumor and the possibility of Yrigoyen's retirement continued to circulate into 1928. *El Dia* (Jujuy), 23 January 1928, published news that Yrigoyen had offered a deal to Alvear in which he would renounce his candidacy if Melo and Gallo did the same.
15. Though quite partisan to Hipólito Yrigoyen and the *UCR*, Gabriel del Mazo, *El radicalismo* provides the best history of the Radical Civic Union. David Rock, *Politics in Argentina* is the best account in the English language.

16. This institutional separation, which had roots dating to 1895, occurred in the aftermath of a boycott by Yrigoyen's faithful of President Alvear's opening address to Congress in May 1924. Gabriel del Mazo, *El radicalismo*, 2: 14–27. Throughout the article I choose to capitalize Personalist and Antipersonalist to highlight the organizational autonomy of these parties. The terms were not capitalized in Argentina at the time. Scholars differ in their approach. For example, Sandra McGee Deutsch, *Counterrevolution in Argentina, 1900–1932* (Lincoln: University of Nebraska Press, 1986) capitalizes the terms, but Richard Walter does not in *The Province of Buenos Aires and Argentine Politics, 1912–1943* (London: Cambridge University Press, 1985).
17. Meeting in early April 1927, the "Convention of the Parties of the Right" (*Convención de los Partidos de la Derecha*) composed of conservative representatives from 9 of the 14 provinces agreed to support whomever won the Antipersonalist candidacy marking the formal start of a "United Front." Archivo General de la Nación, Buenos Aires, Archivo de Julio A. Roca Jr., Legajo 6/116.
18. Carlos Ibarguren, an influential conservative involved in the United Front, wrote that the coalition's enthusiasm was so strong that triumph seemed inevitable. *La historia que he vivido*, 3d. ed. (Buenos Aires: Dictio, 1977), 491.
19. Alan Lascano, *Yrigoyenismo y antipersonalismo* (Buenos Aires: Centro Editor de América Latina, 1985), 64.
20. Félix Luna, *Yrigoyen* (Buenos Aires: Desarrollo, 1964): 379–381.
21. The Personalist Nominating Convention convened on March 22, 1928 in the Opera Theatre of Buenos Aires.
22. Most elections between 1912 and 1930 averaged a voter turnout rate of 60 percent. The 1928 rate was 81 percent. In the presidential elections of 1916 and 1922 respectively, 63 percent and 55 percent of eligible voters participated. These statistics come from *Memorias del Ministerio del Interior* as cited by Darío Cantón in *A Model of Social Change in Latin America* (Buenos Aires: Editorial del Instituto, 1967), 17.
23. Manuel Gálvez underscores these elements of his organizing method, *Vida de Hipólito Yrigoyen*, 369.
24. Articles 5 and 6 in the Constitution sanction the president and/or Congress with the power to intervene a province with autocratic authority over local officials "to guarantee the enjoyment ... and exercise of its institutions." Almost always controversial, federal intervention has often been a political tool.
25. Carl Solberg, *Oil and Nationalism in Argentina*, 127.
26. Luis Huergo, *Memorandum del Petróleo, 8 de abril 1913* (Buenos Aires, 1913), 28-29. Huergo was a civil engineer hired by the Federal Government as a consultant. Huergo's statement echoes anti-United States sentiment circulating in certain quarters to explain the carnage underway in Mexico at the time.
27. The best overview of petroleum development and its political ramifications is Solberg, *Oil and Nationalism in Argentina*. See chapter 1 for the early debates

over nationalization. Arturo Frondizi, *Petróleo y política* (Buenos Aires: Raigal, 1955) is excellent in providing legislative details of debate.

28. Standard Oil agents negotiated for leases in Comodoro Rivadavia in 1913 but were summarily rebuffed. Arturo Frondizi, *Petróleo y política*, (Buenos Aires: Raigal, 1955), 62.
29. Enrique Mosconi, YPF director from 1919 through 1931, chronicles Standard Oil's acquisitions in detail, *YPF contra Standard Oil* (Buenos Aires: Agepe, 1958).
30. Subsoil rights were adjudicated according to the 1886 Mining Code, which provided provincial jurisdiction except in the sparsely populated national territories south of Mendoza and Buenos Aires provinces. Solberg, *Oil and Nationalism*, 14–15.
31. President Alvear exhibited his disinclination to interfere in provincial politics almost immediately upon entering office, when in an address inaugurating the 1923 Congress he determined not to intervene in Córdoba even though the lower house had voted for such action. Del Mazo, *El radicalismo*, 2: 40–41; Alvear's purpose seems to have been to counter the intervention policy of his predecessor, Hipólito Yrigoyen. His determination to shun federal intervention turned into a centerpiece of his presidential administration. See Manuel Goldstraj, *Años y erores: Un cuarto siglo de política argentina* (Buenos Aires: Sophos, 1957), 47–54. Nevertheless, petroleum transcended that concern, and Alvear expressed his nationalist objectives in a message dispatched to Congress in September 1923. Cámara de Diputados, *Diario de sesiones*, vol. 6 (Sept. 20, 1923), 245.
32. The director of the Argentine state oil enterprise *YPF*, Enrique Mosconi, devoted large quantities of time and energy constructing the evidence and the argument for nationalization. Mosconi wrote various books in the effort including *YPF contra la Standard Oil* (Buenos Aires, 1958).
33. Throughout the period of Radical control of the executive branch (1916–1930) various attempts to form a national conservative party were made, but none succeeded. Conservative parties remained organized at the provincial level only. Therefore, I have chosen not to capitalize the word conservative since most provincial parties of that ideological view gave themselves other titles. See Roberto Azaretto, *Historia de las Fuerzas Conservadoras* (Buenos Aires: Centro Editor de América Latina, 1983), 58–70.
34. Details of the murder and subsequent events were reported in a series of articles printed in a provincial newspaper *La Voz del Norte* (Salta). The content of the wanted poster appeared on July 15, 1926. It read, "The Standard Oil Company S.A. of Argentina offers five thousand pesos per head, dead or alive, for the capture of the assassins of señores Hettman, Theisner, and Hidalgo. Signed Juan B. Eskesen, Vicepresident of the Standard Oil Co. S.A."
35. Killed on May 2, 1926, Aníbal Mario's face was so destroyed by gunshots that he was not identified until records of his fingerprints returned from Buenos Aires two weeks later. *La Voz del Norte*, 3 May, and 2 June 1926.

36. Announcement of the release of José Vélez, Enrique Herrera, Nicanor Ojeda, Tránsito Riso, and Guillermo Luna came in *El Intransigente* (Salta), 9 September 1928. The son of the dead French scientist wrote an article that appeared in *La Voz del Norte*, 13 September 1928, reviewing the horror of the case.
37. An extended account of the Senillosa episode and its cover-up may be found in Nicholas Biddle, "Oil and Democracy in Argentina, 1916–1930," (Ph.D. diss., Duke University, 1991), ch. 6.
38. Both were leaders in an organization called the Continental Alliance, which dedicated itself to defending economic resources against foreign exploitation. Argentine oil nationalization was their primary goal. Raul Larra, *El General Baldrich y la defensa del petróleo argentino* (Buenos Aires: Mariano Moreno, 1981), 54-60.
39. This phrase and the above quote are to found in the text of General Baldrich's speech, reprinted in Enrique Mosconi, *YPF contra la Standard Oil* (Buenos Aires: Agepe, 1958), pp. 193 and 222.
40. *La Voz del Norte*, 5 August 1927, reported that General Baldrich had completed his fifth radio conference and that he was planning a trip to Salta. The same paper drew the conclusion quoted on 17 August 1927.
41. Carl Solberg provides a concise description of the bills, designated Dispatch No. 77 and Dispatch No. 95, and the debate in *Oil and Nationalism*, 124–126.
42. Arturo Frondizi, *Petróleo y política* (Buenos Aires: Raigal, 1955), 195–198; Solberg, *Oil and Nationalism*, 124.
43. Eduardo M. Gonella, director of the Seminar of Economics and Finance, used these words in a conference held on August 18, 1927. *La economía del petróleo nacional y su legislación* (Buenos Aires: Imprenta La Universidad, 1927), 4.
44. Cámara de Diputados, *Diario de sesiones,* vol. 3 (18 August 1927), 867.
45. Joseph S. Tulchin reviews the source and consequence of the meat ban in *Argentina and the United States: A Conflicted Relationship* (Boston: Twayne Publishers, 1990).
46. The final bill is in Cámara de Diputados, *Diario de sesiones* vol. 4 (1 September 1927), 361–362. Discussion and voting on the Personalist amendment is in Cámara de Diputados, *Diario de sesiones*, vol. 4 (8 September 1927), 478.
47. Senators in Argentina's bicameral legislature were elected by their provincial parliaments for terms of nine years. At no point throughout the period of Radical Party dominance at the executive level did conservatives lose a majority in the Senate.
48. Socialist Deputy Enrique Dickman accused the Personalists of proffering their program "at this juncture to ready the thirsty fields for the impending presidential election." Cámara de Diputados, *Diario de sesiones*, vol. 4 (19 August 1927), 28.
49. The quote in this sentence comes from Rock, *Politics in Argentina*, 240. The two quotes in the following sentence are taken from Solberg, *Oil and Nationalism*, 127; and Walter, *The Province of Buenos Aires and Argentine Politics*, 80.
50. Luna, *Yrigoyen*, 379-80; Lascano, *Yrigoyenismo y antipersonalismo*, 43 and 62–65.
51. Salta's election date was December 4, followed by Tucumán on January 15, Santa Fe on February 5 and, finally, Córdoba on March 11.

52. Yrigoyen made this comment to a reporter for *Nueva Epoca*, 6 September 1927.
53. Rock, *Politics in Argentina*, 232.
54. The arrival of the delegation was covered in *La Voz del Norte*, 21 September 1927.
55. Molinari periodically suggested Yrigoyen's retirement through 1927 into 1928 in the context of running for president in his place. *El Intransigente*, 27 January 1927; *El Dia* (Jujuy), 23 January 1928.
56. *La Voz del Norte*, 21 September 1927.
57. "How much the times have changed and with the result that the Radical governor creates tight connections among the men who in times past were such enemies . . ." *El Intransigente*, 14 October 1923.
58. Those who welcomed Standard Oil's expansion were, for the most part, large landowners who received lease payments from the firm for exploration and production on their lands. These same individuals composed the bulk of Provincial Union leadership. In August 1925 Güemes vetoed a Standard Oil petition passed by a conservative majority in the legislature to double the size of lands leased for exploration. Mosconi, *YPF contra la Standard*, 122–123; Biddle, "Oil and Democracy," 153–164.
59. Directors of Salta's Center for Radical Workers (*Centro Obrero Radical)* issued a passionate statement denouncing Güemes and the Personalist National Committee for usurping control over local decisions. *El Intransigente*, 26 July 1927. Yrigoyen was reported paying Güemes a personal visit to press upon him "the necessity that he occupy that place of battle." *La Voz del Norte*, 14 June 1927.
60. The text of Güemes's resignation of October 29 was printed in *La Voz del Norte*, 3 November 1927.
61. *La Voz del Norte*, 1 and 5 November 1927.
62. A member of the Provincial Union since the first decade of the century, Julio Cornejo switched parties midway through 1926. Poking fun at the turnaround, the Antipersonalist press asked, "How can this man be a sincere Yrigoyenista?" *El Intransigente*, 27 February 1927.
63. In September 1926 Cornejo demanded that Governor Corbalán release all data concerning the territorial extensions of Standard Oil in the province. *La Voz del Norte*, 9 September 1926. Cornejo voiced continuing allegations and entreaties in September 1926, January, February, March, and May 1927. The complete text of his most recent attack was published in *La Voz del Norte*, 10 August 1927.
64. "The provincialists are not hiding their contentment . . ." *Nueva Epoca*, 4 November 1927.
65. Such practices had been commonplace until Yrigoyen sent federal troops into the provinces to guarantee free suffrage and fair elections in the early years of his first presidential term. Most historians dismiss Yrigoyen's 20 provincial interventions during his first presidency as a mark of corruption. Rock describes these as the imposition of "corrupt client regimes on the interior provinces," *Politics in Argentina*, 115. Though often ordered on flimsy grounds that divulged ulterior motives, Yrigoyen's interventions were also the singular means to enforce fair elections, as was the case in Salta in 1918. Biddle, "Oil and Democracy," 30–37.

66. One journalist described that election as follows: "Never before has the pressure of the patrons on their farmhands and renters been felt with more rigor; forced custody, written and marked ballots, threats, seductions, the purchase of votes, all have been put into practice in order to circumvent freedom of suffrage and the secret ballot." *El Intransigente*, 3 January 1925. An extensive list of electoral violations was published by *El Intransigente*, 3 December 1924.
67. The Antipersonalist announcement in favor of Güemes and an expression of disgust at the possibility of colluding with conservatives in support of Alvarado appeared in *El Intransigente*, 21 October 1927.
68. Manuel Alvarado was nominated by conservatives on July 20, 1927. The Antipersonalist press scorned the candidate two days later. *El Intransigente*, 22 July 1927.
69. Alvarado lost to Joaquín Castellanos in the December 1918 election culminating Yrigoyen's first federal intervention in Salta. *Nueva Epoca*, 16–21 December 1918. For details of the Alvarado reactionary stand against the Sáenz Peña Law and universal suffrage, see Biddle, "Oil and Democracy," ch. 2.
70. *El Intransigente*, 21 October 1927.
71. *Nueva Epoca*, 31 October and 21 November 1927, reported how the conservative governor, Joaquín Corbalán, pressured public employees to participate in party functions; *Nueva Epoca*, 10 November 1927, described the arrest of Radicals hanging posters; *La Voz del Norte*, 23 and 27 November 1927, related the destruction of party headquarters by both police and Provincial Union members; *Nueva Epoca*, 7 November 1927, narrates the beating of a Radical in the presence of a sheriff and a provincial legislator; *La Voz del Norte*, 27 November 1927, told of constant threats on Radicals' lives by police and also reprints a telegram sent to Yrigoyen and the National Committee by Carlos Borzani listing all these various offenses.
72. *La Prensa* was the largest daily in the nation through the period and while its editors liked to consider the paper above partisan politics, they showed themselves routinely partial to the conservative perspective. *La Prensa* (Buenos Aires) 13 November 1927, p. 13; and 14 November 1927, p. 11.
73. *La Prensa*, 22 November 1927, p. 13. Robustiano Patrón Costas headed the Provincial Union for roughly twenty-five years until his forced retirement in the wake of the June 1943 coup d'etat leading to the rise of Juan Perón. Still little known historiographically, Patrón Costas is credited by Solberg as having been "the arbiter of Argentine politics" through the 1930s, *Oil and Nationalism*, 160. He was, as well, a central architect of both the United Front in 1927 and the Concordancia in 1931 that undergirded the openly corrupt regimes of that decade. Biddle, "Oil and Democracy," chs. 5–7.
74. Benjamin Villafañe and Matías Sánchez Sorondo, *La palabra de un patriota sobre el problema de la legislación del petróleo* (Buenos Aires: Imprenta Dominguez, 1927), 30 and 39. The book is a compendium of the addresses given by the two at a conference held at the Instituto Popular de Conferencias, Buenos Aires, in November 1927. Villafañe adds an introduction. He wrote

prolifically publishing two other books in 1926 and 1927 entitled, *Yrigoyen no es un partido político* and *Yrigoyen, el último dictador*.

75. While there is no evidence that Matías Sánchez Sorondo participated directly in the conspiracy to overthrow Yrigoyen, he was aware of it and supportive. Furthermore, Sánchez Sorondo became the first interior minister in the provisional government of General Uriburu. Walter, *The Province of Buenos Aires and Argentine Politics*, ch. 3.
76. Fenelon Arias wrote a firsthand account of the events that were published three days after their occurrence in *La Voz del Norte*, 23 November 1927.
77. Accounts of the rampage in Orán ran four consecutive days in *La Voz del Norte*, 20–24 November 1927. The editorial in *La Prensa* appeared 23 November 1927, p. 13.
78. This is the general perspective taken historiographically to explain the passage and implementation of the Sánez Peña Laws. See Rock, *Politics in Argentina*, 39–41.
79. *La Prensa*, 25 November 1927, p. 15.
80. *La Prensa*, 4 December 1927, p. 15.
81. This was the unlikely and unfounded explanation reported in *La Prensa*, 4 December 1927, p. 15.
82. The Personalist charge accompanied the story describing Carlos Borzani's message to the interior minister on December 3, "Confirmed in all parts my information about interrupting the telegraph lines as part of the plan which the government will execute the day of the election." Printed in *La Voz del Norte*, 4 December 1927.
83. A flurry of accounts reporting fraud and coercion were published in *La Voz del Norte*, 6 December 1927. The most outlandish incident happened at a poll officiated by Senator Luis Linares of the Provincial Union and Luis Langou, a local Radical officer. According to Langou, at one point during the day Senator Linares walked into the enclosed voting booth where the ballot box was located. After a few minutes, Langou followed him to see what the matter was, only to find the conservative senator pilfering the box and ripping up votes cast for Radicals. Through the following week various newspapers detailed the episode and roundly scorned Senator Linares, but no legal action was taken. Linares's action gained the attention of *El Intransigente*, *Nueva Epoca*, and *La Epoca* in the days and weeks ahead.
84. Official results of the voting were not known for a month, which was only a little longer than the average period necessary to transport ballot boxes from remote districts, validate their contents and count the votes. The complexity and cumbersomeness of compiling results is described in detail in the extensive report compiled by Manuel Carlés during his tenure as Yrigoyen's "interventor" in the province in 1918. Manuel Carlés, *Intervención Naciónal en Salta: Informe elevado al Ministerio del Interior por el Interventor Naciónal Manuel Carlés* (Buenos Aires: Imprenta Talleres Gráficos, 1919).
85. Reprinted in *La Voz del Norte*, 12 December 1927.

86. *La Epoca*, 10 December 1927, reprinted in *La Voz del Norte*, 11 December 1927.
87. *La Voz del Norte*, 29 December 1927.
88. Federico Pinedo described the failure of conservative victory in these terms in an interview conducted by Luis Alberto Romero, *Federico Pinedo: Una historia oral* (Buenos Aires: Instituto Torcuato di Tella, 1971), 6. Gubernatorial elections in Salta utilized an electoral college that held 53 votes in 1927. Three candidates ran in the 1927 race, one of whom, Benjamin Zorilla, garnered enough popular ballots to control four electoral college votes. Though in years past a member of the *Provincial Union*, Zorilla turned against the party in the wake of the Sencillosa scandal and, therefore, threw his four electoral college votes to the Radical Julio Cornejo. These votes were just enough to hand Cornejo the victory. For more details of Salta's gubernatorial electoral procedures, see Nicholas Biddle, "Oil and Democracy," pp. 282–297.
89. Luna puts the margin of Cornejo's victory at 200 votes, *Yrigoyen*, 381. The number of voters registered for the 1928 presidential election was 40,295, *La Voz del Norte*, 14 April 1928.
90. *Crítica* (Buenos Aires), 6 January 1928, reprinted in *El Intransigente*, 10 January 1928.
91. *La Voz del Norte*, 10 January 1928.
92. *Crítica*, 10 January 1928.
93. *La Prensa*, 13 January 1928, p. 14.
94. *Nueva Epoca*, 20 March 1919. This organ of Salta's Provincial Union published numerous attacks against the Sáenz Peña Law for a decade after its passage.
95. Carlos Serrey, *En el parlamento y fuera de el* (Buenos Aires: Editorial El Ateneo, 1944), 255. Carlos Serrey spoke these words on the Senate floor in June 1928 in his capacity as Provincial Union Senator from Salta.
96. The Provincial Union boycotted the Provincial Electoral Convention charged with approving the vote and seating the new governor. This defiance forced a federal intervention in late April. *El Intransigente*, 14 and 19 April 1928.
97. *La Voz del Norte*, 22 April 1927.
98. Gabriel del Mazo, *La primera presidencia de Yrigoyen* (Buenos Aires: Centro Editor de America Latina, 1986), 40.
99. *La Prensa*, 30 November 1927, p. 15, carried the story of ten armed men roaming countryside terrorizing the "tranquil inhabitants" in the interests of the Radical Party.
100. *La Prensa*, 1 January 1928, p. 13, reported the death of a conservative at the hands of local police in Tucumán, news of which reached President Alvear in a formal denunciation (*denuncia*) that petitioned for federal intervention.
101. This overly polite phrase to describe political opponents was used in *La Prensa*, 14 January 1928, p. 13. Accounts of the brawls that became routine at political events may be found daily in *La Prensa* from 2 through 14 January 1928.
102. *La Prensa*, 12 January 1928, p. 16, describes the arrival in Tucumán of five groups from different political parties and provinces. The same article contains

the quoted headline and an account of two violent eruptions, one at an Antipersonalist rally and another in front of a Personalist campaign office. Police intervened effectively in the first case, but four people were shot in the second.

103. *La Prensa,* 14 January 1928, p. 15, contains descriptions of the brawl and shooting as well as the Interior Minister's statement.
104. *Nueva Epoca,* 23 March 1927, reported Alvear as having made the comment during an interview with Julio Roca and Rodolfo Moreno, the organizers of the Convention of the Parties of the Right.
105. The most intimate rendering of Marcelo Alvear and his presidential administration is found in an account written by a close friend, Manuel Goldstraj, *Años y erores: Un cuarto siglo de política argentina* (Buenos Aires: Sophos, 1957). Felix Luna, *Alvear* (Buenos Aires: Sudamericana, 1988) is informative but simultaneously condescending.
106. *La Prensa* 2 February 1928, p. 15.
107. Luis Alberto Romero, "Federico Pinedo: an oral history," *Instituto Torcuato de Tella* (Buenos Aires, 1971), 6. Pinedo recalls that Antonio Rubirosa, the United Front's treasurer, made the comment to him in January 1928.
108. According to Gálvez, some in the minister's inner circles were also still loyal to Yrigoyen and they exposed the conspiracy at an incipient stage. *Vida de Yrigoyen,* 370.
109. The denial was published in *La Nación,* 21 February 1928. Though he disputed being "capable of using arms to create dictators," in hindsight we know General Justo to have been deceiving himself, if not also the public since, of course, in September 1930 he and General Uriburu led Yrigoyen's overthrow. Gálvez notes that in February 1928 "the start of the conspiracy [culminating in September 1930] was discovered." Gálvez, *Vida de Yrigoyen,* 368–371. Luna provides a more terse but very similar account in *Yrigoyen,* 382. See also Robert A. Potash, *The Army and Politics in Argentina, 1928-1945* (Stanford: Stanford University Press, 1969), 18–21.
110. Ibarguren freely describes in general terms the early scheming in 1927 between ultra-nationalists and General Uriburu to overthrow Yrigoyen. *La historia que he vivido,* 520. Julio Irazusta discusses early conversations with General Uriburu that were interrupted because the latter felt it proper to wait for his imminent retirement before involving himself in what was, objectively, treason. Julio Irazusta, *El pensamiento político nacional,* 2 vols. (Buenos Aires: Obligado, 1975), 2: 11–13. *La Prensa,* 1–15 March 1928, contains daily descriptions of electoral violence that often resulted in death.
111. After the Personalist victory in Santa Fe, non-allied Antipersonalists published a declaration excoriating "the invasion of our hard-working province by organized bands of undesirable elements at the margin of the law." *La Prensa,* 26 February 1928, p. 12.
112. For the genesis of the *Liga Republicana* see Sandra McGee Deutsch, "The Right under Radicalism, 1916–1930," in McGee Deutsch and Dolkart, eds., *The Argentine Right* (Wilmington, Delaware: Scholarly Resources, 1993). Rock

quotes a proclamation of the *Liga Republicana* in October 1929 that attacks the validity of elections and representative democracy wholesale. *Politics in Argentina*, 249–251.

113. *La Prensa*, 9 March 1928, p. 14.
114. A conservative partisan identified only as Señor Viñas leveled this accusation at a Conservative Party convention in La Plata, Buenos Aires. *La Prensa*, 16 March 1928, p. 17.
115. *La Prensa*, 19 March 1928, p. 12.
116. Luna, *Yrigoyen*, 385.
117. Felix Luna, Roberto Etchepareborda, and Gabriel del Mazo tend to subscribe almost literally to Yrigoyen's magical powers, although the phrase quoted is from *La Epoca* as reproduced in Carl Solberg, *Oil Nationalism*, 126. Manuel Gálvez, and more appreciatively, David Rock and Richard Walter find the Radical political machinery in the Littoral to be Yrigoyen's most fundamental bulwark of success. Carl Solberg emphasizes oil as the primary factor to the 1928 election.
118. Solberg, *Oil Nationalism*, 127.
119. Buenos Aires and Tucumán were the only two provinces with Personalist governors. Clearly Buenos Aires was the centerpiece of the political map, however, it is instructive to note that Yrigoyen received 56 percent of his votes from outside both the province and the city of Buenos Aires. Roberto Etchepareborda, *La crisis del 1930* (Buenos Aires: Centro Editor de América Latina, 1983), 38 and 40.
120. From Betschtedt's report to the Personalist National Committee reprinted in *La Voz del Norte*, 12 December 1927.
121. *La Prensa*, 9 March 1928, p. 14.
122. Rock describes interior provinces as having political environments "where elections were still largely a matter of winning the support of the local *hacendado*, who could browbeat the peasants into voting whichever way he wished." *Politics in Argentina*, 62. The emphasis on the progress of Argentine politics from 1912–1930 is underscored by the regress of the "Infamous Decade" that followed from 1930–1943. For a particularly nostalgic view see J.J. Hernández Arregui, *La formación de la conciencia nacional*, 290–293.
123. The subtle nature of these public relations is reflected in U.S. Ambassador Robert Bliss's communication to the State Department two months prior to the election. "I find," he wrote, "among sober-minded Argentines … pride in Argentina's having gone so long without a revolution," which assured the Ambassador that a second Yrigoyen presidency would be accepted by some as "a lesser evil to a coup d'etat and the more serious events it is capable of bringing in its trail." [i.e. production and trade disruptions] Robert Bliss to the Secretarty of State, 8 February 1928, United States National Archives, Department of State, Record Group 59, 835.00/418.
124. Peter Smith also argues that the 1930 coup d'etat resulted from a crisis of legitimacy that followed from the pre-1912 crisis in participation. Smith, however,

blames the Radicals for continuing "to act as an opposition party once they were in power," while I suggest that Radicals had no choice since conservatives treated Personalists less like a ruling party than as enemies of the state. Smith, *Failure of Democracy: Conflict among Political Elites, 1904–1955*, (Madison: University of Wisconsin Press, 1974): 96–97; Biddle, "Oil and Democracy," pp. 327–336.

125. *La Prensa*, 26 February 1928, p. 11, predicted that current electoral conditions would turn voters "in favor of the party that presents the greatest guarantees for the public wellbeing. . . ."
126. The appearance of the *Klan Radical* and the *Liga Republicana* within 12 months were the most dramatic manifestations of unrestrained political confrontation and violence. See Sandra McGee Deutsch, "The Right under Radicalism, 1916-1930," in Sandra McGee Deutsch and Ronald Dolkart, eds., *The Argentine Right*, (Wilmington, Delaware: Scholarly Resources, 1993).
127. Gálvez, *Vida de Yrigoyen*, 375.

CHAPTER FOUR

The Organization of Jujuy's Sugar *Ingenios* in a Regional Context (1870–1940)

Marcelo Lagos

Introduction

Argentina's sugar industry has its origins in the country's northwest in the seventeenth century, but it was only in the final third of the nineteenth century when it grew to such an extent that it quickly became the economic engine of that region. The industry's development in this regional context is a complex one, and it is impossible to understand the history of this important economic sector of the Argentine interior without taking into account the social, economic, and political changes that caused the development of the sugar industry.[1] This chapter offers a new analysis of the industry's history in the province of Jujuy from these multiple perspectives, highlighting similarities and differences with the industry's history in other parts of the country.

There have been three principal areas where the Argentine sugar industry has prospered, each corresponding to chronological stages in the industry's history in the country: Tucumán-Santiago del Estero, Salta-Jujuy, and the Littoral. Tucumán pioneered the industry in Argentina and never lost its position as the industry leader, despite periodic crises and ups and downs. The contemporary experience of nearby Santiago del Estero was brief, and the industry failed to reach sufficient levels of production that would allow it to compete with its small provincial neighbor, Tucumán. The Salta-Jujuy industry, located in the subtropical, woodland strip known as the *tucumano-oranense,* or the *yungas,* grew together at a later stage but achieved in three of their *ingenios* (plantations combining fields and mills) a level of development and concentration that came to equal and then surpass the individual production of any of the Tucumán *ingenios.* The third region and stage in the industry's history took place primarily in Formosa, Chaco, and Corrientes, with some development also in Misiones and in northern Santa Fe. It never contributed a significant amount to the industry's total production, nor did it play the leading role in these provincial economies that it did in the previously mentioned provinces.

The geographic focus of this analysis is the eastern part of Jujuy (see map IV), Argentina's northernmost province, and specifically the San Francisco river valley, an area where three sugar *ingenios* (again, a term that encompasses both the agricultural and industrial facets of sugar production) emerged whose development transformed the region. This is a subtropical region located in the Andean foothills at an elevation varying between 400 and 600 meters and bordering on the Chaco woodlands. Its hot climate and fertile lands were exceptionally suitable to the cultivation of sugar cane. However, the region's rainfall—abundant in the summer but almost nonexistent in the winter—was not sufficient to meet the needs of the sugar plantations and made necessary the use of river irrigation. The need to control water resources was one of the factors that led the *ingenios* to monopolize land in the zone. A network of artificial irrigation, which between major and feeder canals covered thousands of kilometers, was the response taken by the *ingenios* in order to ensure the only resource that nature had not amply supplied the region and that would allow the maximum growth of the sugar cane.

The primary purpose of this chapter is to analyze the development of the sugar agro-industry and the impact it had on the enclave microregion and the provincial economy. In order to accomplish this, it is also necessary to analyze certain related questions that will allow the reader a fuller understanding of sugar's history in this part of the Argentine interior. First of all, some attention must be given to the conditions that existed when the industry first appeared. Second, the move from the primitive productive practices of the region's sugar haciendas to the incorporation of technology and the expansion of the industry is also an important question. The conditions surrounding the insertion of the local industry in the regional market and then in the national one are other variables to consider. It also is necessary to give some attention to the evolution of the local labor market whose nexus is the sugar industry and will eventually link other regions to the sugar-producing centers. Finally, the political factor, that is to say, the influence on the provincial government that the powerful sugar companies were acquiring, needs to be discussed.

The Emergence of the *Ingenio*-Plantation

Eastern Jujuy was for almost three centuries a frontier region. The fact that it bordered on the Chaco, a territory dominated by seminomadic indigenous peoples until the first decade of the twentieth century, influenced enormously both its landholding patterns and the character of its society. An area of forts, garrisons, missions, and vast haciendas, the region's scarce population reflected a frontier society: soldiers, creoles employed on the haciendas, and indigenous groups displaying different degrees of integration into the dominant culture comprised the small but heterogeneous core of its population. Until the middle of the nineteenth century, the isolated and dispersed haciendas of the zone lived essentially from cattle raising and the cultivation and rudimentary production of sugar cane, refined to produce alcohol and supply the Jujuy market, especially the indigenous population found in the province's highlands.

Roads were extremely primitive and only passable during drought months. Cities were nonexistent. San Pedro and Ledesma (today the most important cities, after the capital, in the province) did not even qualify then as towns. The population was scattered, with important concentrations only being found around the haciendas. The first important regional changes were associated precisely with the transformation of the two largest haciendas in the region, San Pedro and Ledesma, into important sugar firms with great concentrations of capital investment and technology. That process, which takes place between 1870 and the first years of the twentieth century, was characterized by changes in landholding patterns, the incorporation of capital, machinery, technical personnel, and a labor force drawn from outside the region. This process broke down the region's isolation, a fact that permitted it to channel production toward other provincial as well as national markets.

Out of the process of incorporating technology and capital there emerged what can be called the *ingenio*-plantation. These productive units combined two features: agricultural property producing primary material in large quantities, and the modern factory that carried out all the stages in the manufacturing of sugar. Slowly, and in keeping with the rate at which the market expanded, the *ingenios*' functions came to include as well the control of the commercialization of sugar, resulting in a high degree of concentration. The concentration of activities was one of the distinctive traits that since their origins differentiated the Jujuy sugar firms from the Tucumán industry.

A two-stage chronological distinction can be employed to characterize the evolution of the *ingenio*-plantation. The first can be called the "take-off" stage, beginning in the 1870s when the initial investments in technology were made, the new factory establishments began production, and the first harvests of high yield were gathered. This stage ends around 1915, when the local firms found themselves in conditions to compete with others in the industry and win a greater share of the national market. The second stage runs roughly from 1915 to 1940, and might be called the stage of "insertion and definitive consolidation" in the national market by virtue of being a period of a great boom and expansion for the industry, despite passing through the crisis of the 1930s. Of course this periodization is far from being exact, and in the years indicated as turning points there did not occur dramatic changes or abrupt ruptures; they serve only to roughly indicate the beginning and end of stages that did as a whole represent something different. The differences between the periods encompass many facets and include everything from the political influence that the companies had at the provincial and national level to the composition of the labor force. However, it is necessary to make clear that there are continuities for both stages, as well as features whose origins are found in the first stage but only come to fruition in the second. This temporal demarcation, with whatever arbitrary elements it may have, will serve as a framework both to facilitate understanding of the history of the industry and to differentiate the steps in the same process.

The take-off stage was characterized by a number of early difficulties. One of the principal ones was the isolation of the sugar-producing region where the enclave firms were to be found. During the first 30 years of their history, the *ingenios* only could get their product out of the region via poor departmental roads to the nearest station of the *Ferrocarril Central Norte.* The Perico station, some 40 kilometers from San Pedro and 90 from Ledesma, received the carts loaded with sugar or alcohol and then had to ship them to the rest of the country. The problem of transportation was thus vital for the development of the agro-industrialists of the zone.

The Leach brothers, owners of "La Esperanza" *ingenio,* saw the matter clearly and thought that a water route via the Bermejo River was possible, despite previous failed attempts.[2] In 1899, they undertook an expedition from Lavayen to the city of Corrientes. At issue was a study of the possibilities for trade with the Littoral and eventually the Rio de la Plata, epicenter of the national consumer market.[3] However, the expedition revealed that the river was not navigable a good part of the year, nor was it possible to transport large volumes of goods. The idea was temporarily discarded and all hopes were placed on the railroad.

An official circular from the ministry of public works in 1901 emphasized the interest of the sugar interests in the arrival of the railroad:

> It is necessary to add that landowners and industrialists in the region which the railroad line crosses have offered a free donation of lands they occupy, as well as donations of wood and other construction materials, loans of money redeemable through freight rates, aid that is estimated at a million pesos.[4]

In 1905, a trunk line, "El Ramal," a name that would soon be given to the entire region, arrived in Ledesma. Contact with the outside world was established and the possibility of lowering costs through reduced freights was now a fact. With the transport difficulties overcome, the time to the consumer markets became shorter.

Investment of Domestic and Foreign Capital: The Formation of Corporations

Modernizing productive capacities depended on the technology applied that, in turn, depended on the availability of capital. The contribution of the latter followed a familiar story. Initially, particularly in the latter years of the hacienda era and the first of the sugar mills, the capital came from local sources in Salta and Jujuy. To these were added foreign capital, capital that arrived above all during the stage of technological improvement and the expansion of the sugar estates. A subsequent stage witnessed the formation of corporations with important investments by finance capital from both Buenos Aires and foreign sources.

The "La Esperanza" mill can be analyzed as a case study in the history of the sugar industry in Jujuy. Around the middle of the nineteenth century, Juan N. Fernández Cornejo began to cultivate sugar on the San Pedro hacienda, a hacienda that had belonged until then to the family of Salta businessman, Martín Otero. The Cornejos were a Salta family who had immigrated in the eighteenth century to Argentina from Peru, the country from which they had brought sugar cane for planting. Indeed, they had a long relationship with the industry, having founded around 1760 the oldest sugar mill in northern Argentina, the "San Isidro en Campo Santo" mill located in Salta.

After the first endeavors in San Pedro, Miguel Francisco Araoz, brother-in-law of Juan N. Fernández Cornejo, took over the business, establishing in

1882 the firm, "Araoz, Ugarriza, Uriburu and Company," with the participation of Rogelio Leach, the direct descendent of the British family soon to play an important role in the local industry. Years later, Leach himself would lease to his partners the entire "La Esperanza" *ingenio* while remaining in partnership with them in certain important properties such as "San Pedro," "Pichanal," and "Pampitas." In 1888, the "Araoz and Leach Co." was established with Rogelio, his five brothers, and their Salta-born offspring serving as partners. In 1893, the firm, "Leach Bros." was established with a large capital investment. The "Leach Bros. Inc." was subsequently formed with the participation of Francisco Leach and his four brothers (Rogelio had died).

The Leach corporation's capital comprised some $300,000 pesos, with each partner contributing an equal fifth, a figure that rose to $600,000 in the final stock subscription in November 1915.[5] To have an idea of what these figures meant, we need simply look at the total revenue of the province of Jujuy, which in the year of the firm's establishment was $278,670 pesos, that is to say, it did not even reach the company's initial capital value; and by 1915, it was still a mere $1,455,790, that is only double of what the British brothers had deposited in banks in Salta and Buenos Aires.[6] Subsequently, "in the year 1912 was established in London ... the Leach Argentine Estates Ltd., with a capital of 1,052,500 pounds sterling, divided between preferred, ordinary, and deferred stock."[7]

The region's other two *ingenios* would have a similar history. These *ingenios* belonged to old established families such as the Ovejeros and the Alvarados, and early in the century associated with French capital ("Ledesma") and German capital ("La Mendieta"). At a subsequent stage, these *ingenios* were also incorporated. In 1909, after a series of setbacks, the "Ingenio La Mendieta S.A." was established with its legal residence in Buenos Aires and capital assets of 1 million pesos. The principal stockholders were Messrs. Arning and Hapsberg.[8] In 1914, the Ledesma *ingenio* reached its final metamorphosis with the establishment of the "Ledesma Sugar Estates and Refining Company Ltd."

Land Acquisition and Consolidation of the Enclave Economy

Capital was needed not only for the introduction of technology into the mills but also to acquire land, especially around the *ingenio's* productive nucleus. One testimony of the period states:

> In the San Pedro and Ledesma districts, it is very difficult to acquire through purchase, land suitable for agriculture given the fact that the owners of the La Esperanza and Ledesma *ingenios* buy those cultivable plots containing water, for purposes of planting sugarcane, at better prices than offered by any other.

It can be said that there do not exist other landowners in the region than the aforementioned.[9]

Between the final two final decades of the nineteenth century and the first of the twentieth, a high percentage of the lands of the low-lying districts and the central valleys (the Capital, San Antonio, and El Carmen districts) began to be monopolized by the *ingenios.* These properties were the heart of the plantations and the mills and had the following dimensions in 1901: La Esperanza 25 square leagues, Ledesma 15 square leagues, and La Mendieta around 5 square leagues. Based on the property appraisal rolls, which unfortunately are incomplete, I have attempted to reconstruct the evolution of the *ingenios'* property in the two districts where they were based. In the first column of the three tables below figure the period in which the appraisal was made, in the second the name of the owner or the legal name under which the companies operated, in the third the appraisal in pesos, in the fourth what this amount represented with respect to total properties assessed in the district, and the last the total number of properties with an assessment (see Tables 4.1–4.3).

Table 4.1 Evolution of the Estate Lands of the Ledesma *ingenio* (1872–1919), Ledesma District

Year	*Owner*	*Value (in Pesos)*	*Percent of District Appraisals*	*Total Properties in Ledesma District*
1872	Ovejero Hermanos	100,000	39.8%	6*
1878	Ovejero Hermanos	100,000	37.1%	22
1883	Ovejero Hermanos	150,000	38.5%	33
1887	Ovejero Hermanos	200,000	32.7%	70**
1891–1894	Ovejero and Grande	350,000	34.2%	77
1895–1899	Ovejero and Zerda	360,000	34.2%	146
1901–1903***	Ovejero and Zerda	500,000	44.4%	135
1904–1908	Cia Azuc. Ledesma	610,000	48.7%	200
1910–1915	Cia Azuc. Ledesma	1,300,000	figures incomplete	130
1915–1919	Ledesma Sugar	6,866,800	figures incomplete	92

* If we include in the figures Salvador Villar, also a planter of sugarcane, two landowners between themselves monopolized 79.6 percent of the appraised properties.
** Five landowners monopolized 71.8 percent of the appraised properties.
*** From 1901 on, urban and rural properties were appraised separately. The above figures and those that follow consider only rural properties.
Sources: Archivo Histórico de Jujuy. Registro de Catastro de la Provincia de Jujuy. Years: 1872, 1878, 1883, 1887, 1891, 1894, 1895, 1899, 1901, 1904, 1908, 1910, 1915, 1919.

Table 4.2 Evolution of the Estate Lands of La Esperanza *ingenio* (1872–1919), San Pedro District

Year	*Owner*	*Value (in Pesos)*	*Percent of District Appraisals*	*Total Properties in San Pedro District*
1872	M. Araoz	35,000	49.2%	28*
1878	M. Araoz	40,000	43.0%	30
1883	M. Araoz	50,000	34.0%	52
1887	Araoz, Uriburu, and Cornejo	310,000	39.7%	90
1891–1894	Araoz Hnos./Araoz and Leach	125,000 368,500	figures incomplete	192
1895–1899	Araoz and Leach/ Leach Hnos	132,500 235,600	figures incomplete	205
1900–1903**	Leach Hnos	651,500 150,000***	47.7%/13.3%	170
1904–1908	Leach Hnos and	715,200 210,000***	72.7%/16.7%	281
1910–1915	Leach Arg. Estates	9,530,200 3,611,000***	figures incomplete	335
1915–1919	Leach Arg. Estates	6,724,100***	77.3%	183

* Until 1899, landowners in the Santa Barbara district are included.
** From 1901 on, owners of urban and rural property are listed separately.
*** Appraisals of the properties located in the Ledesma district. The most important estates are San Lorenzo, Campo Colorado, and Sauzal.
Sources: Archivo Histórico de Jujuy. Registro de Catastro de la Provincia de Jujuy. Years: 1872, 1878, 1883, 1887, 1891, 1894, 1895, 1899, 1901, 1904, 1908, 1910, 1915, 1919.

Table 4.3 Evolution of the Estate Lands of La Mendieta *Ingenio* (1891–1919), San Pedro District

Year	*Owner*	*Value (in Pesos)*	*Percent of District Appraisals*	*Total Properties in San Pedro District*
1891–1894	Justino Alvarado	10,000	figures incomplete	192
1895–1899	Faustino Alvarado	30,000	figures incomplete	205
1901–1903	Alvarado and Co.	272,000	21.7%	170
1904–1908	La Mendieta S.A.	380,000	includes properties from other districts	281
1910	La Mendieta S.A.	475,000	figures incomplete	–
1915	La Mendieta S.A.	–	figures incomplete	335
1915-	La Mendieta S.A.	1,625,000	18.6%*	183

* If the Leach estates are added to this figure, both *ingenios* together possessed 95.9 percent of the appraised properties in the district
Sources: Archivo Histórico de Jujuy. Registro de Catastro de la Provincia de Jujuy. Years: 1872, 1878, 1883, 1887, 1891, 1894, 1895, 1899, 1901, 1904, 1908, 1910, 1915, 1919.

I am aware of the problems posed by these appraisals. Their trustworthiness and accuracy are questionable since there did not exist in these years precise appraisal standards, nor did the state have a single criterion on how to actually assess them. There also was lacking reliable personnel to make them, indeed with the landowners themselves frequently responsible for the appraisals of their own real estate. But in a case such as Jujuy's, in which these are the only statistical sources to trace the evolution of land tenancy, and to be sure used in conjunction with qualitative sources (which also reveal the steady increase in the *ingenios'* holdings), there is no other recourse but to try and make some use of them. The appraisals demonstrate that toward the end of the period under analysis, in the 1930s, the total properties of the firms were greater in size than the very boundaries of the districts in which they were located. Once again, the Leach family can serve as a point of reference. While the San Pedro district, core of the Leach's *ingenio,* comprised some 186,000 hectares, the British company's lands covered 192,000 hectares.[10]

I have not been able to find in the province's land appraisal registry whether Jujuy's *ingenios* acquired land in areas removed from the *ingenio*-plantation nucleus. Nevertheless, I believe I am in a position to refute the hypothesis of British scholar, Ian Rutledge, with respect to the supposed direct relationship between appropriation of the land and the acquisition of a labor force, a hypothesis based on his study of San Martín Tabacal (an important *ingenio* established in 1919 located in the Orán district in Salta province), whose experience he believes possibly could have broader applicability.[11] For the period under study, my research indicates that there was no "satellitezation" by the Jujuy firms of areas removed from their estates for purposes of acquiring a resident labor force. The lands were acquired more for purposes of production and the diversification of economic activities than to ensure a captive labor force. Jujuy's *ingenios* had been operating for 40 years when San Martín del Tabacal was created and had mechanisms for assuring themselves a steady labor force. The monopolization of land distant from the *ingenios* was perhaps an extreme measure that Jujuy's *ingenios* did not need to resort to by virtue of already having an assured labor supply and having other mechanisms of labor recruitment.

The tendency to hoard land went hand in hand with the monopolization of all aspects of production that took place within the estates. Testimonies from the period offer us the image of the *ingenios* as "states within states." The existence of palisades controlling the entrance to the properties, the banning of the movement of people within the estates not authorized by the companies, the control exercised over all means of communication within them, the very establishment of the place of residence for the estate's population, reveals firms disposed to exercise direct control over their domain. It should not surprise us that contemporaries, specifically those who condemned this situation, considered the *ingenios* to be true "feudal fiefdoms."

My utilization of the terminology of the era (which appears repeatedly in political speeches and in the opposition press, both Radical and the Socialist) does not mean in any way I adhere to those historical studies that see in the monopolization of land a phenomenon tending to reproduce social relations of a "feudal" nature. On the contrary, from the economic rationalism that compelled the acquisition of land, as well the capitalist logic with which such undertakings were run, to the nature of labor recruitment, control, and the distribution of the labor force, not to mention the market destination of what the estates produced, the inappropriateness of the categorization of "feudal" is obvious. The loose use of the term does not, however, mitigate the fact that there did occur an unquestionable hoarding of land accompanied by a pronounced tendency toward absolute control by the firms, control over relations of a strictly economic nature to those of a more social kind. A single, graphic example of this was that the firms ended up monopolizing and exercising powers generally the preserve of the state. The police, the justices of the peace, and municipal government all fell under their influence. All were more directly dependent on the *ingenios'* administration than on orders emanating from the distant and supine provincial governments.[12]

Economic Diversification

In addition to a greater concentration of capital, land, and political power, the *ingenios* experienced in these years a process of economic diversification (see below diagram [figure 4.1]). Although sugar production constituted the primary source of profits and growth, the *ingenios* (especially Ledesma and La Esperanza) devoted capital and labor to a wide variety of activities; some were of an experimental kind with an unpredictable future, others had a bountiful nature at their disposal (and therefore were less risky). It is well known that the *ingenio*-plantation tended toward self-sufficiency, depending to the minimum degree possible on inputs from outside its own borders. For that reason, it established around its central nucleus (administration-*ingenio*) veritable islands of complementary economies that fulfilled the triple function of maintenance of the central plant, maintenance of the permanent and transitory personnel, and selling in regional and national markets.

The below diagram is an idealized scheme of the complementary activities to sugar developed by Jujuy's *ingenios* in the first two decades of the

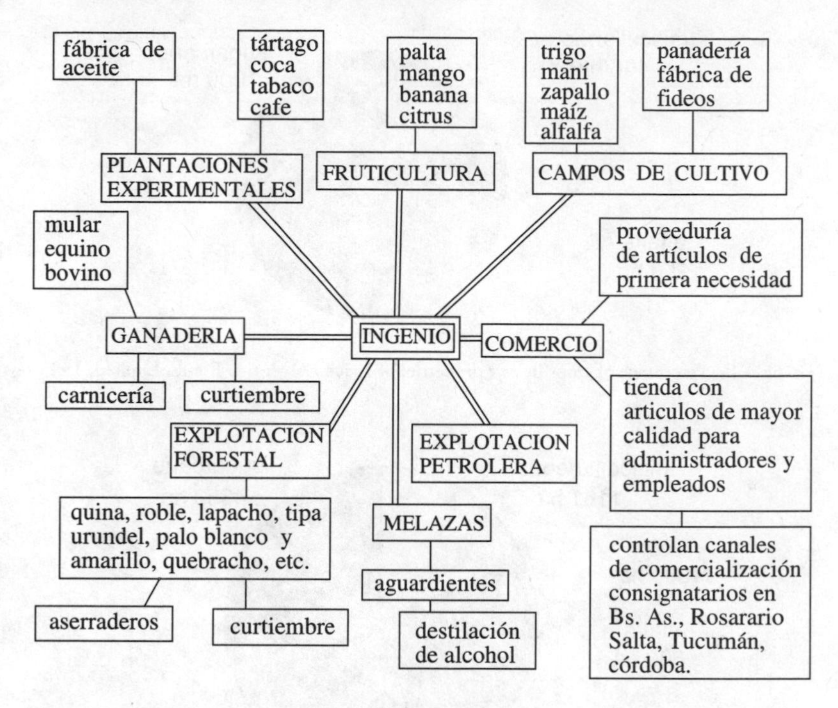

Figure 4.1 Complimentary economies to the Ingenio-Plantation

century. It should be emphasized that not all of the *ingenios* achieved the same level of diversification (it was less in La Mendieta and greater in La Esperanza). The purpose of the diagram is simply to emphasize the wide range of activities undertaken by the firms. From the diagram it can be surmised that, except for tools and machinery, all the necessary inputs were produced by the *ingenio,* especially those related to the material needs of the labor force. We do not have the sources necessary to determine with exactness the importance that each one of these complementary activities had with respect to the total productive unit. The figures tend to be precise when they deal with sugar, vague with regard to other undertakings. For purposes of determining the distribution of the crops in the two most important *ingenios,* below are diagrams (graph 4.1 and 4.2) that indicate the situation until around 1920. These are the San Lorenzo and Campo Grande properties belonging to the Leach family, and the San Antonio and Ledesma estates, owned by the Ledesma *ingenio.*

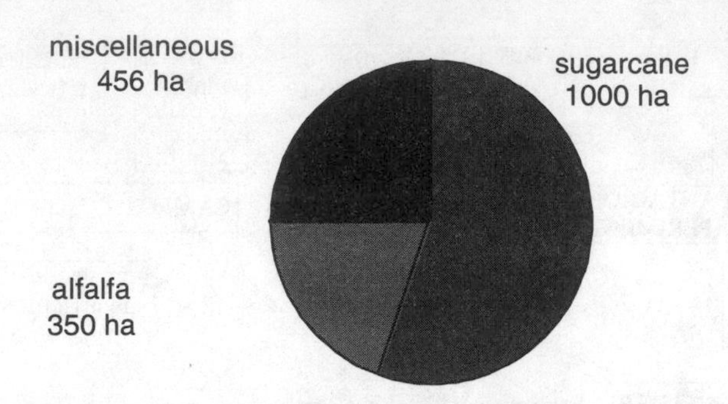

Graph 4.1 Percentage of crops in two properties of Leach's Argentine Estates Limited, 1923–1927

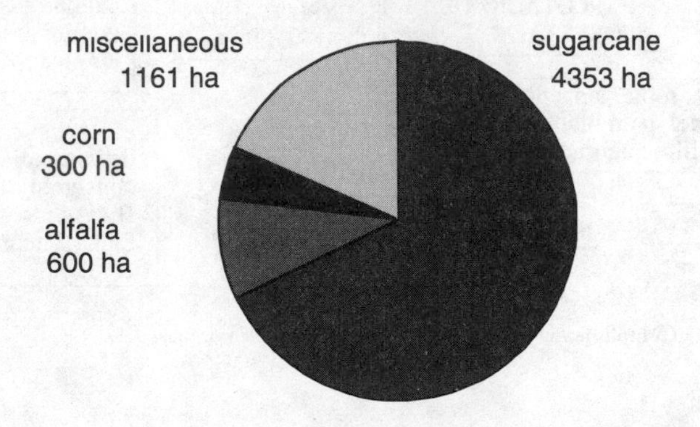

Graph 4.2 Crop percentages in two properties of Ledesma Sugar Estates, 1923–1927
Source: Catastro general de la propiedad teritorial, pcia, de Jujuy, 1923–1927

Despite the fact that sugarcane constituted without question the most important crop, it is interesting to note the importance of the other crops, especially those in the "miscellaneous" category. In the case of the Leach properties, these were primarily coffee, coca, and bananas, and in the Ledesma estates citrus products especially. Also noteworthy is the amount of uncultivated land on these estates, 75 percent in the case of Ledesma and 98 percent on the Leach family lands. The uncultivated lands were "without irrigation, uncultivated, mountainous, with scrub, swamps," but from which were extracted the wood that fueled the boilers and that was sold in the market; at the same time, they served as pasture land for livestock, with some 5,000 head on the Ledesma properties and 3,000 in San Lorenzo, the great majority being cattle and to a lesser degree goats and horses.

We see in the economic diversification another distinctive characteristic of the Jujuy *ingenios* compared to those in Tucumán. None of the latter managed to encompass such a wide variety of productive activities as those in Jujuy, being strictly monoculture sugar producers; their owners devoted some of their profits to other activities, but generally they did so by sending their capital outside the province to other regions such as the Chaco, where they invested in both *quebracho* and cotton cultivation.[13] Obviously, the availability of suitable land, added to the natural conditions unique to the region, allowed the Jujuy firms to undertake diverse ventures. Although important as a whole, it is also necessary to remember their essentially "complimentary" nature to a sugar economy, which never ceased to be the leading economic sector of the region.

Insertion in the National Market

The decade stretching from 1910 to 1920 witnesses the beginning of serious competition between the Jujuy and Tucumán *ingenios.* The amount of land under cultivation in Jujuy rose from 3,200 hectares in 1910 to 10,900 hectares in 1915, finally declining around 1920 to what would be its stable level of 10,000 hectares (see graph 4.3). These figures demonstrate that while Tucumán's output was stagnant, the Jujuy firms not only were increasing the amount of land under cultivation but were achieving an ever great output per hectare. Graph 4.4 shows the steady growth of Jujuy's three *ingenios.*

The combination of a series of factors allowed the local industry to penetrate the national market, a market at moments saturated with Tucumán's sugar. The combined influences of a series of bad harvests (1906, 1907, and subsequent ones) with critical moments of overproduction (1913, 1914), added to the exhaustion of the native sugar cane, made this a period of crisis for the Tucumán industry. It also became clear that an industrial elite

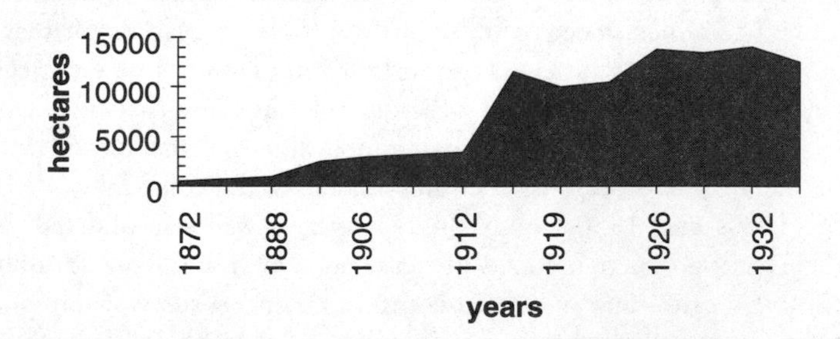

Graph 4.3 Land devoted to sugarcane cultivation in the province of Jujuy, 1872–1934

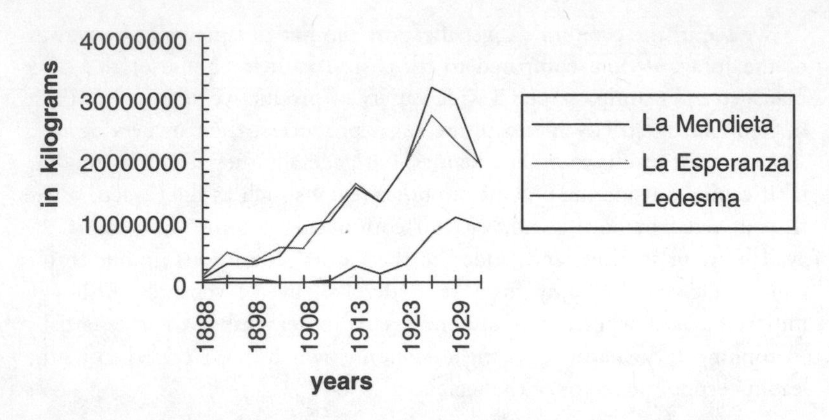

Graph 4.4 Sugar production in the Jujuy *ingenios*, 1888–1933

(generally owners of small establishments in the case of Tucumán) independent from the producers of the primary material was less able to adapt to upheavals in the market than were those who had emerged with a high degree of concentration, as in the case of Jujuy.

Thus there was absent in Jujuy's sugar industry "the antagonistic duality and complementarity of *latifundio* and *minifundio,*" and "the division between agricultural and industrial sectors" that gave rise in the Tucumán case to the independent cane farmer. The concentration of land and technology gave Jujuy's industry a more typically Latin American character than was the case in Tucumán. Peru's Pacific coast *ingenios* and the Brazilian *usinas* abound in similarities to Jujuy's industry.[14] As one observer noted, "Jujuy's sugar region presents an economic and social organization very different from that of Tucumán. Here there are no independent cane farmers; every productive center is a vast fiefdom isolated in the middle of the jungle."[15] The nonexistence in Jujuy of the middle agrarian sector that was the independent cane farmer is demonstrated in graph 4.5 where it can be seen in the La Mendieta *ingenio*—the only one in Jujuy that did not mill only its own sugar cane—the small proportion of cane acquired from producers in the San Pedro district independent of the company.

The 1920s and 1930s found Jujuy's *ingenios* well consolidated. They were able to set production quotas in consultation with the Tucumán industry, establishing a cartel that prevented competition and ensured steady profits. Such agreements revealed that the insertion of the Jujuy and Salta *ingenios* in the national market was increasingly important. Nonetheless, it is necessary

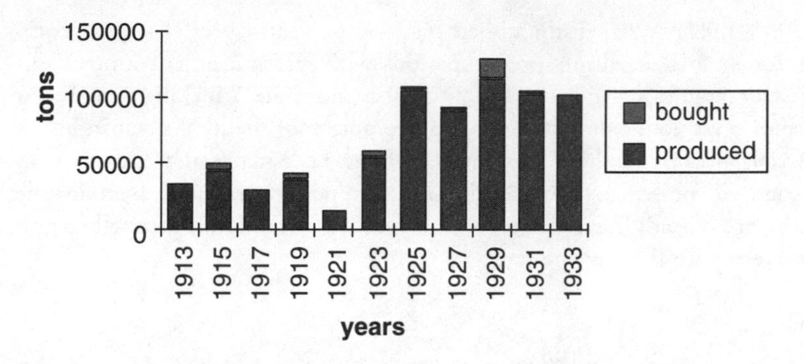

Graph 4.5 Proportion of sugar production of the La Mendieta Ingenio 1913–1923

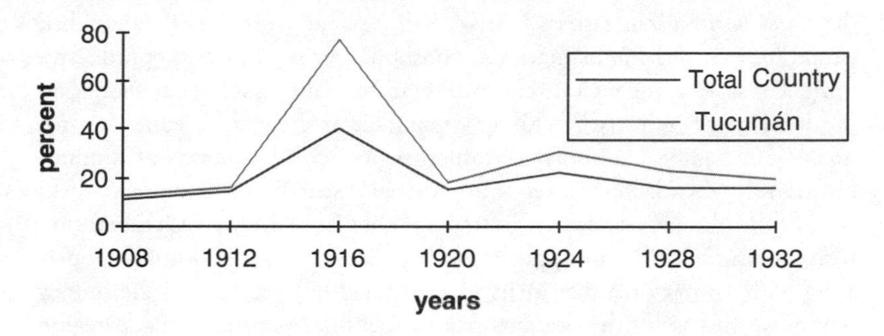

Graph 4.6 Percentage of Jujuy's sugar production with respect to that of Tucumán and country as a whole, 1908–1932

to avoid exaggerating the size of Jujuy's agro-industry. It is enough simply to compare the proportion of Jujuy's production in relation to Tucumán's and its percentage of the country's sugar production as a whole to appreciate its real dimensions (see graph 4.6).

However, with respect to the province, sugar by now constituted the principal economic activity and the very existence of the provincial state would have been threatened by its decline or demise. Benjamín Villafañe and Herminio Arrieta typified the new era: respectively, the politician more concerned about protecting and fomenting the local sugar industry than in public affairs and the businessman who made politics an extension of his private economic power.[16] Villafañe, an anti-Personalist Radical, served as provincial congressman (1909–1910, 1917–1918), national congressman (1920–1924), governor of the province (1924–1927), and finally senator

(1932–1941). What distinguished his long political career was less a commitment to Radicalism's preoccupations with reforming the country's oligarchic politics or widening the franchise and more a lifelong crusade for regional economic development and promotion of the local sugar industry. Arrieta became in these years owner of the Ledesma estates and used his economic power to become the most important political boss in the province. Clearly there was a confusion of the sugar firms' private business interests with those of the state.

Formation of the Labor Market

The formation of a regional labor market begins with the transformation of the sugar *haciendas* in eastern Jujuy toward the end of the 1870s, though it would be the *ingenio*-plantations' consolidation as the only productive centers of scale in the region that would create a labor market dependent on the local sugar industry.[17] This labor market emerged in a zone that for many years had been an isolated frontier region with few means of communication. It was a labor market with a demand strictly for temporary workers (during the harvest season lasting roughly from May to October) and with an insufficient local supply of workers. It was also a labor market operating with a workforce that in its majority had not yet entered the money economy, the case of the peasants from Argentina's *puneño* highland region and from Bolivia, as well as the indigenous workers from the Chaco who were still members of hunting-gathering societies. There emerged therefore a labor market with some pronounced peculiarities with respect to a so-called "free labor market."

Because they were unable to meet their needs with the local labor supply, the sugar firms ended up devising different strategies to create a flexible workforce, integrating various microregions into their productive complexes for purposes of labor supply. These microregions were marginal areas inhabited either by indigenous peoples with economies disrupted by the advance of the State and white society or by peasants who, for reasons of a diverse nature, were passing through crises in their traditional subsistence economies. Thus there took place around the *ingenios* a permanent "to-and fro" movement that had as regions of labor expulsion the poorest parts of northern Argentina—the *puneña* zone, the Calchaquies valleys, and the Chaco forests—and as the focal point of temporary resettlement, the sugarcane plantations of eastern Jujuy. If the integration of this work force into the labor market was incomplete, it was due to the capitalistic logic of the firms that benefited from a situation in which the social reproduction of these people was their own concern whenever their labor was not needed by the companies.

It is not, however, accurate to speak about the existence of a free labor market during the "take-off" stage in Jujuy's sugar industry. The procurement of the labor force, the relations of production, and especially the forms of compensation were of an extremely heterogeneous nature; but the predominant characteristic was various forms of coercion, both of a physical and legal nature, over the labor force. It might appear to be a paradox that these agro-industries, which from their origins had a clearly capitalistic nature, would create around them a labor market with "pre-capitalist" (in the broad sense of the term) features. This really is not such a contradiction, since firms do not necessarily reproduce capitalist labor markets if these are not to the benefit of their profitability, indeed they can reinforce archaic practices, perpetuating them through the law itself.

The free labor market that did gradually take shape during the period under analysis was the product of a slow process that closely mirrored the development of the agro-industry and only acquired this character of "free" once the ready availability of a labor force (which only occurs in the 1930s) made it possible for the businessmen to deepen capitalist relations with workers. Thus, for example, the procurement of labor was achieved both through coercion and money incentives, varying according to period and the kind of worker to be contracted. Coercion is much greater in the first years of the take-off stage and was exercised with greater force against the indigenous laborer. Coercion took the form of both forced levies of Indian labor, especially of the various Chaco tribes such as the Matacos and Tobas, as well as encroachment on subsistence economies of these same tribes by the Army followed by permanent settlers. Monetary incentives increased in importance with time until becoming the principal factor behind the incorporation of labor, and was the greatest pull on impoverished peasants who began to see the sugar harvest as a complement to their own harried subsistence economies.[18]

Payment of the labor force followed a similar process. There was a general tendency to gradually abandon payment in kind, vouchers, the retention of salaries, and running accounts, in favor of regular payments in the national currency. But this was a very slow transformation that in reality occurred in a hybrid fashion. As late as the 1930s, we find, along with a more widespread monetary relationship, the persistence of payment in vouchers, the "tarja," and account book deductions.[19] The sugar companies always had under their control the handling and regulation of relations with the labor force. That absolute control was favored by a combination of factors. In the first place, the relative isolation and near absolute domination of the sugar firms in the region allowed them to exercise a rather arbitrary control of the resident population there. To this must be added the existence of legislation that regulated labor relations between owners and employees favoring the former. At the same time, the nonexistence of trade union organizations to fight for the workers' rights strengthened the owners' control. The seasonal nature of the work and the high degree of illiteracy among the labor force undoubtedly were two important factors that explain the delay in achieving any kind of trade union organization (labor unions would not become a reality in the region until the Peronist period). But we must also take into account the zeal with which the *ingenios'* administrators watched over workers who sought to make demands for improvements in their living and work conditions.

Indeed, the state and the law collaborated in support of the *ingenios* at the moment of the formation of the regional labor market. For the provincial government and the local ruling class, the end-of-the-century stage in the sugar industry's development marked a moment of transition and transformation. Hoping not to stay on the sidelines during the process of modernization, the local ruling class understood that the agro-industries of the *Ramal* would constitute the driving axis of the regional economy. The overlapping of interests attained with alacrity commercial, political, and also family dimensions. Ideologically identified with the growth of the agro-industries and increasingly dependent on the financial power of the *ingenios,* the provincial state was making way for new groups of power.

Throughout the period under study, the administration and internal workings of the *ingenios* was organized around units of production in which the croplands were divided in order to rationalize production. But there exists a distinction with respect to their administration in the two periods we have noted in the industry's history. The first was characterized by a more direct and centralized control, the second by a greater involvement of contractors and agents, that is, intermediaries responsible to the administration for the operation of the *ingenios.* The role of estate administration did not change greatly, but the importance being assumed by the intermediaries in managing the labor force did, intermediaries who acted as labor contractors, or *conchabadores.* It was to the *ingenio*'s advantage and proved to be more efficient to set aside funds for purposes of relinquishing direct control of the labor force to such intermediaries and thus part of the responsibility for anything having to do with contracts and labor relations in general.[20]

With the Leach estates again serving as an example, we can see the nature of labor relations reflected in the *ingenio*'s very organization. The spatial distribution of the estates reproduced the plan of the central administration around the Pueblo Ingenio, or the "manor" of the Leach family. The manor and administrative offices were the focal point and heart of the *ingenio*-plantation, to approach the latter was to encounter a powerful symbol, since it represented authority within the company.[21] Around the administration were found the houses of the professional, technical, and administrative staff. These were already brick structures by the beginning of the century and had running water and electricity that, as with the other utilities, were provided and controlled by the company. A bit more removed could be found the settlements, shacks, and huts where there lived the so-called *caseños* or the permanent personnel, employees engaged both in work in the processing plants and work in the fields. Finally, a good distance from owners place of residence, were found the dwellings of the seasonal workers, those who remained six months in the *ingenio,* as well as the sites set aside so that the aborigines from the Chaco could pitch their "huetes."[22]

The Provincial State and the Ingenios

During the period under analysis, political power that was initially in the hands of traditional sectors tied economically to livestock, commerce, and land ownership was retreating before the economic power of the *ingenios.* Financial support from the *ingenio* owners made it possible to control the machinery of provincial government and establish politicians in the service of the sugar firms, especially in that part of the governmental machinery related to the direct interests of the *ingenios:* taxation, control of the labor market, water rights, etc. An increase in political power was paralleled by the consolidation of the economic power of the *ingenios.* The decade of the 1930s represented the culmination of the fusion of economic and political power, the years when the identification between the two was absolute and, with the province firmly under control, the local sugar interests began to move into the national arena.

The stanchions of the *ingenios*'s advance in provincial politics were essentially two: taxes and credit. From the late nineteenth century, the *ingenios* went on to become one of the economic sectors that provided the greatest amount of revenue to the provincial government (graph 4.7). If around 1890, taxes on sugarcane, refined sugar, and stills represented only 3.8 percent of the province's fiscal resources, in 1893 they already constituted 8.2 percent and in 1896, 11.8 percent, finally to reach 25.8 percent in 1899. At the beginning of the century, the percentages increased significantly. After descending to 14.1 percent in 1906 they rose to 22.2 percent in 1909, 32.1 percent

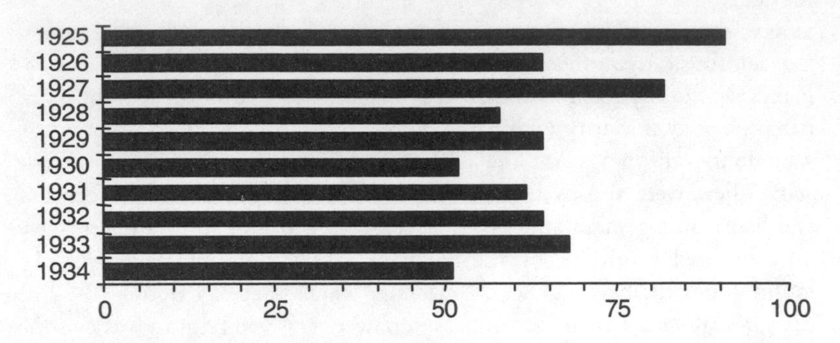

Graph 4.7 Percentage of public fiscal resources coming from taxes on sugar, alcohol, and derivatives, Jujuy province, 1925–1934

in 1912 and 52.5 percent in 1915.[23] In 1925,they reached a high point of contributing 90 percent of the provincial budget, thereafter to decline but to hold steady for the remainder of the decade at a figure oscillating between 50 and 60 percent.[24]

Through lending policies, the *ingenios* became the provincial governments' principal creditors. Loans were made possible by high interest rates paid to the firms, with the government generally making payments in the form of reducing the tax burden on the *ingenios.* The first documented loans were made in 1901. One for $50,000 pesos was made by Ovejero y Zerda, owners of Ledesma, for the construction of the municipal market[25]; another for $22,000 was handed over to the provincial treasury by the Leach Brothers and Co.[26] It is easy imagine the influence the sugar interests wielded by being the government's creditors. In 1914, a loan was negotiated with the province's three *ingenios.* One of the provisions of the loan stipulated that the provincial government could not raise the taxes on the sugar firms for a period of ten years.[28] Subsequent Radical administrations accepted loans on identical terms, thereby hobbling the government's tax policies. We find important loans in 1921 and 1922 made during the administration of Mateo Córdova and another sizable one during Benjamín Villafañe's government.[28]

The influence of the sugar interests on the state widened when their representatives actually began to occupy government posts, especially in the legislature. In that regard, the 1930s stand, as stated earlier, as the most representative period of the *ingenios'* power. An analysis of the legislature's composition in 1932 eloquently supports this assertion. Of the 18 members of the legislature, 17 belonged to the *Partido Popular* (representing the Conservatives). Only one, Rodolfo Aparicio, was a member of the Radical Party. Of the 17 conservative representatives, the great majority had direct ties to one of the three *ingenios:* Luis De Santis belonged since 1917 to administrative staff of the Leach Brothers, and in 1932 was the *ingenio*'s administrative head; Juan Martín Sylvester had been a doctor at La Esperanza since 1926 and was subsequently director of the San Pedro hospital; Francisco Leach, a native *salteño,* worked on his father's estate and was a large landowner in El Carmen, the district he represented in the legislature; Luis María Oliver, another Leach relation, was assistant manager of the company's warehouses; Mamerto Salazar, a justice of the peace for various years in Yavi, was a landowner of plantations in Tarija and manager of the Yavi plantation owned by the Patrón Costas family, in addition to working as a labor contractor for Ledesma; Carlos Undiano was a doctor at the La Mendieta *ingenio;* Ramón Outon, associated for years with the Ledesma *ingenio,* was in charge of the *ingenio*'s administration for some time; Alberto Caracciolo was a merchant from Ledesma closely tied to company interests; Lázaro Taglioli, one of the most important Ledesma contractors, had as his zone of influence the *puna* and southern Bolivia; Antonio Vargas Orellana, *ingenio* contractor, held various municipal offices in the Ledesma department. Other members of the legislature, such as Daniel González Pérez, Arturo Pérez Alisedo, Fernando Berghmans, and Pedro Buitrago, went on to become governors of the province in the 1930. Three of these individuals were lawyers whose firms counted among their clients the *ingenios.*

Conclusion

The development of sugar agro-industry in eastern Jujuy unleashed profound changes that not only transformed this peripheral frontier region into the epicenter of the province's principal economic activity, but through its productive undertakings managed to insert Jujuy for the first time in its history into the national consumer market. If the subtropical region had lacked great weight in the province's economic and social development until well beyond the first half of the nineteenth century, the transformation of rudimentary haciendas into powerful *ingenios* marked a turning point in the economic balance of power, making it the richest and most dynamic region in the province, a region where besides generating jobs, enough wealth was created to allow the state to collect taxes to meet new needs. The region was also transformed in environmental terms. The planting of sugarcane led to the clearing process (*desmonte*) and the original vegetation of the jungle that covered the valleys and lowlands disappeared, with what remained of difficult access.

In the enclave economy of the microregion, all changes were intimately tied to the development of the agro-industry. The growth of the population, its composition, the urbanization process, the layout of the railroad system, the very organization of the government bureaucracy, were the fruit of the conjunctural necessities of the take-off and consolidation of the sugar industry. This was not necessarily a completely positive process. The interests of the region became identical to those of the companies, and economic development was not an integrated process but one that encompassed only that which caused the agro-industry to function and grow. Through their ownership of vast properties, the *ingenios* for all practical purposes become owners of the departments they were located in, indeed the latter being regarded as their private property. The local aristocracy's logic associated the companies' growth with the region's progress, and thus no restraints were imposed on their activities, not even those that were naturally the government's prerogatives. The relationship between the provincial government and the local aristocracy with the *ingenios* became more complex as the importance of the latter increased, eventually leading to social bonds and intertwined concrete economic interests. The overlapping connections rapidly established commercial, political, and also family ties. Increasingly dependent for its finances on the economic power of the sugar industry, the state would continue to make room for new groups of power. In the final two decades of the period under analysis, the identification of the region's economy with the *ingenios* had reached such extremes that the very provincial state could be included as part of the agro-industry's empire.

Once the first steps of the sugar industry had been taken in Jujuy, the *ingenios* concentrated on four broad objectives to foment their development: monopolizing land to ensure themselves the vital resource of water and to prevent the emergence of new competitors; the incorporation of technology that would allow for reducing production costs; fomenting transportation improvements that would put the industry in more direct contact with the consumer market; and the establishment of a labor market that would serve the firms' needs for an irregular supply of labor. These four objectives all were geared toward a final objective: to penetrate the national consumer market, especially in the Littoral provinces. To achieve that end, they would have to compete with the powerful and established competitor that was the Tucumán sugar industry. Relations with the latter were conflict-ridden and competitive and for years lacked a forum to work toward their peaceful resolution in the principal business organizations representing the industry, controlled by Tucumán until the 1920s, at which point Jujuy's industry was finally strong enough to make its voice heard.

The crucial step that was the penetration into the domestic market was due to a series of factors of both an external and internal nature. The external factors were those stemming from the crisis of the Tucumán industry itself and the window of opportunity that the Jujuy *ingenios* were able to capitalize on. The internal factors were those of achieving good harvests, obtaining greater output per hectare, the refining of their own sugar cane, the establishment of a transportation network to the consumer markets, and creation of a labor market more economically efficient than Tucumán's. None of these factors, however, was as important as the high degree of concentration achieved by Jujuy's *ingenios.* Such concentration gave them, compared to their rival, greater stability in the face of market fluctuations and limited the frictions between refiners, planters, and workers, since the sugar firms controlled all facets of production: the plantation, the mills, even distribution.

Translated by James P. Brennan

Notes

1. On the history of the sugar industry in the Argentine north, see Donna Guy, *Política azucarera argentina: Tucumán y la generación del 80* (Tucumán: Ed. Fundación Banco Comercial del Norte, 1981); Marcos Giménez Zapiola, "El interior argentino y el desarrollo hacia afuera: el caso de Tucumán," *El Régimen oligárquico* (Buenos Aires: Amorrotu, 1975); Jorge Balán, "Una cuestión regional en Argentina: burguesías provinciales y mercado nacional en el desarrollo agroexportador," *Desarrollo Económico,* 18: 69 (1978); and Daniel Santamaria, *Azúcar y sociedad en el Noroeste argentino* (Buenos Aires: Ed. del IDES, 1986).
2. G. Araoz, *Navegación del Río Bermejo y viajes al Gran Chaco* (Buenos Aires: Ed. Imp. Europa, 1884).
3. Francisco Clunie, "La comunicación fluvial entre el Chaco Occidental y el Río Paraguay. Navegación del Bermejo por los Srs. Leach," *Boletín del Instituto Geográfico Argentino*, XX, (1899).
4. Pablo Balduin, "Historia de San Pedro," Jujuy: (unpublished manuscript, 1987): 106.
5. Archivo de los Tribunales de Jujuy, Legajo 8, Año 1935, Expte 156 Testimony of Francisco Leach.
6. Both figures come from Salomé Boto de Calderari and Marcela Whienhausen, "Dinámica y estructura del ingreso y gasto público jujeño, 1890–1915," Jujuy (unpublished manuscript, 1988).
7. Juan de Borja, *Album biográfico e histórico de Jujuy* (Jujuy: Imp. del Colegio, 1934) p. 88.

8. David Moffat, "Informe sobre el Ingenio La Mendieta," Jujuy, (unpublished, 1950): 2.
9. Luis Rodríguez, *La Argentina en 1908* (Buenos Aires: 1908), p. 222.
10. Juan de Borja, *Album biográfico e histórico de Jujuy*, p. 196.
11. Rutledge demonstrated that the Patrón Costas family, one of the most powerful aristocratic families of the Argentine northwest, acquired great amounts of land in the highlands of the *Puna* in Salta and Jujuy for purposes of ensuring themselves the potential labor force they needed. The peasant was obliged to pay his lease through labor services, participating in the annual sugar harvest. The author claims to be uncertain that all the Salta-Jujuy *ingenios* adopted such a strategy but believes it was one of the most common means by which the integration of the highland peasantry into the labor market of the sugar economy was attained. See Ian Rutledge, *Cambio agrario e integración. El desarrollo del capitalismo en Jujuy, 1550–1960* (Tucumán: ECIRA, 1987), p. 187; and Kenneth Duncan and Ian Rutledge, eds., *La tierra y la mano de obra en América Latina* (México: Fondo de Cultura Económica) p. 242.
12. There exists diverse documentation in the Jujuy provincial archive with respect to the naming of policemen, justices of the peace, and other functionaries at the request of the *ingenios,* as well as the arrangements for the payment such public officials were to receive. In the same way, the establishment of the municipal governments offers the spectacle of repeated elections of businessmen and administrative personnel from the industry to important offices.
13. Roberto Pucci, "La élite azucarera y la formación del sector cañero en Tucumán (1880–1920)," in *Conflictos y procesos de la Historia Argentina Contemporánea,* 37 (Buenos Aires: CEAL, 1989).
14. See Manuel Moreno Fraginals, *La historia como arma y otros estudios sobre esclavos, ingenios y plantaciones* (Barcelona: Crítica, 1983); Peter Klarén, "Las consecuencias sociales y económicas de la modernización de la industria azucarera peruana, 1870-1930"; Peter Eisenberg, "Las consecuencias de la modernización para las plantaciones azucareras en Brasil en el siglo XIX"; and Jaime Reis, "Del Bangué a la usina: aspectos sociales de la modernización de la industria azucarera de Pernambuco, Brasil 1850–1920," all in K. Duncan and I. Rutledge, eds., *La tierra y la mano de obra en América Latina.*
15. Pierre Denis, *La valoración del país. La República Argentina 1920* (Buenos Aires: Ed. Solar, 1987).
16. Benjamín Villafañe was governor of Jujuy from 1924 to 1927 and later senator for the province. Herminio Arrieta managed the Ledesma firm in the 1930s and 1940s and was congressman and then senator for Jujuy between 1934 and 1943.
17. This question has been previously studied in Viviana Conti, Marcelo Lagos, and Ana Lagos, "Mano de obra indígena en los ingenios de Jujuy a principios de siglo," in *Conflictos y procesos en la historia argentina contemporánea* 17 (Buenos Aires: CEAL, 1988) and "Conformación del mercado laboral en la etapa de despegue de

los ingenios azucareros jujeños (1880-1920) in *Estudios sobre la historia de la industria azucarera,* ed. Daniel Campi (Tucumán: UNT-UNJ, 1992).

18. For a fuller discussion on these forms of coercion employed in Jujuy's sugar ingenios, see my "Conformación del mercado laboral en la etapa de despegue de los ingenios azucareros jujeños (1880–1930)."
19. The "tarja" was the notebook in which the workers' work assignments and daily rations were recorded.
20. On the various activities and strategies of the intermediaries in contracting and supervising the labor force, see Marcelo Lagos, "Cambios y permanencia del mercado laboral de los ingenios azucareros jujeños en la etapa e inserción al mercado nacional (1920–1940)" in Daniel Campi, *Estudios sobre la historia de la industria azucarera,* vol. III.
21. Scott Witherford, *Workers from the North Plantations, Bolivian Labor and the City in Northwest Argentina* (Austin: University of Texas Press, 1981); Gabriela Karasik, *El control de la mano de obra en un ingenio azucarero. El caso de Ledesma (Pcia de Jujuy)* (Jujuy: ECIRA, 1988).
22. "Huete" is the name of the indigenous huts. Built with sticks, straw, the husks of sugar cane, etc., they had a cone shape. The Indians built these dwellings quickly and with great skill.
23. Salomé Boto and Marcela Whienhausen, op. cit.
24. Figures obtained from Emilio Schleh, op. cit.
25. Archivo Histórico Jujuy, Mensaje del Gobernador Sergio Alvarado a la H. Legislatura, May 1, 1901, Jujuy.
26. Archivo de la Legislatura de Jujuy. Registro Oficial No. 12, Años 1900–1901. Folio 151. Decreto 453. Jujuy, January 16, 1901.
27. Archivo Histórico Jujuy, Año 1914. Caja No. 3, Jujuy, November 15, 1914. Legajillo 11 folios.
28. ALJ, "Libro de Actas," No. 28, Law 502, May 2, 1922, folio 597, "Libro de Actas," No. 28, Law 505, June 2, 1922, folio 314; AHJ, Intervención nacional a cargo de Dr. Carlos Gómez, "Informe sobre el estado financiero de la Provincia al 31 de Diciembre de 1923," Publicación oficial, Jujuy, 1924; ALJ "Libro de leyes y resoluciones de la H. Legislatura de la Provincia" No. 7, 1924-1926, Law 615. 13 de Noviembre de 1924.

CHAPTER FIVE

Forestry in the *Llanos* of La Rioja (1900–1960)

Gabriela Olivera

Introduction

For a number of decades in Argentina, it has been common to regard the national context as a "given," one which begins automatically after independence and the establishment of the central government's political machinery, and to identify the country's history with that of its hegemonic region, the Pampean Littoral. Interest in regional history is relatively recent. Histories about distinct parts of the country are only now being written, histories that will later effectively make up "the national," "the Argentine." Historians of twentieth-century Argentina are now undertaking a reconstruction of the parts, the interregional relations of a world that in the nineteenth century Buenos Aires did not absolutely dominate, and that recognized multiple origins and connections.

This chapter is part of that collective undertaking. It deals with a relatively small geographic region (4,600,000 hectares), one situated in the Argentine northwest and characterized by a depressed agricultural economy and "demographic vacuum," virtually cut off from the process of the formation of national markets until early in the twentieth century. When integration did occur, as was the case with other regions, it happened only after the building of the railroads. Regional history needs to take into account the heterogeneity of those processes of regional insertion in national markets and of the precise historical moments when such processes took place. In that regard, it is appropriate to differentiate the moments of integration for those parts of the northwest whose growing economies complement those of the Pampa and supply the internal market (the case of Tucumán, Salta, and Jujuy's sugar industries) and those that are not initially attractive during the process of national capitalist development (for example, Salta's fruit and tobacco sectors), as well as those that are marginalized by the process and whose principal function is to be suppliers of a labor force for the new regional economies in expansion. In some of these marginal areas, certain things produced in them gain a relative importance (for example, the case of livestock or forestry), not in the sense of complementing the Pampean economy, but rather as also serving as support for other regional economies. The region under study belongs to this latter group.

Not only does the relatively late incorporation into the national markets leave a distinctive mark on the history of this region of the Argentine northwest; the economic sector that permits this process does as well: extractive forestry. This sector constitutes an extractive industry of an itinerant nature, since the boundaries of exploitation move as timber stands are exhausted, something, in turn, that triggers a dynamic of population displacement in the rural sector and has devastating effects on the environment. As part of this process, the solar energy synthesized over the course of centuries in this arid land and transformed into arboreal resources will be extracted and exported to meet the needs of urban centers (wood and charcoal for fuel) and of other regions (trellises for Cuyo's wine industry and fencing posts for the Pampa's agriculture).

This regional analysis will attempt to explain the impact that extraregional ties had on a local economy and society, relying primarily on the sources provided by local archives. To the extent that this glance at the regional dynamic does not restrict itself or the vision of its protagonists to the narrow boundaries of the province, it will have gone beyond a mere provincial chronicle. It is only in this sense, I would argue, that one can intellectually justify regional history. In the following pages I will analyze the economic base of the *obrajeros* (the businessmen who managed a web of economic interests revolving around the forestry business) and how these businessmen sold the forestry products, the manufactured goods (which could be marketed after the arrival of the railroad), and the household-produced goods that until the moment of the forestry industry's expansion had not to any great extent been incorporated into the money economy. I will also explain how a salaried labor force was recruited, the process whereby the timber stands were acquired, the appropriation of land, and the advancement of sectoral interests that without question influenced the process of regional accumulation. To accomplish this objective, it is first necessary to assess the general characteristics of this economic cycle, defining the moments of the extractive industry's "take-off," consolidation, and decline, taking into account as well the organization of the labor process. Finally, I propose to explore by what means the labor force was attracted and subjugated in the extractive sector and the impact on the regional economy that the widespread use of local household production and the existence of seasonal migratory flows had.

Forestry in the "Gran Chaco"

The "Gran Chaco" is a vast subtropical plain extending some 1,090,000 square kilometers wedged between the Andes and the Brazilian frontier, sitting in a great sub-Andean depression that penetrates the interior of three countries: Paraguay, Bolivia, and Argentina. It has been described as a "rural satellite with respect to the most productive forestry and agricultural zones of these countries: the Argentine *Pampa,* eastern Paraguay, and the humid *puna* and *yungas* of Bolivia."[1]

Comprising part of the "Gran Chaco" is the southern Chaco, or the "Chaco Austral," an area covering some 500,000 square kilometers and divided by several rivers into three principal zones: a wet eastern zone (with average annual rainfall more than 1,000 millimeters), a dry western zone (less than 700 millimeters), and an intermediary belt (between 700 and 1,000 millimeters). The region under study is found within the borders of the area of maximum aridity, since its annual rainfall is under 550 millimeters (between 200 and 400 millimeters), a characteristic manifested in the vegetation, with a declining density of trees as ones moves west towards the *Llanos* of La Rioja (which extend to the Andes), until the point that it is practically grasslands.

According to Jorge Morello, the vegetation of the western belt of the Chaco is characteristically dry, subtropical bush. The forest lands in this region are discontinuous and are represented primarily by the *quebracho blanco,* the only common tree of the zone.[2] Bruniard distinguishes two basic forest zones and therefore two forestry economies in the Chaco. One the one hand, there is the eastern zone in which there predominates species of *urunday, lapacho negro,* and *quebracho colorado,* used for the extraction of tannin via industrial processing establishments.[3] On the other, in the western zone, there are found great stands of *quebracho blanco,* whose wood was used for railroad ties, posts and fuel (firewood and charcoal), an activity undertaken by the ranchers of the zone.[4] For his part, Denis argues that these differences in the characteristics of the forest resources and the exploitation of the Chaco's timber stands are associated with two markets: the production of wood in the western zone for the domestic market and in the eastern zone a timber industry intended primarily for foreign markets.[5]

In the "Gran Chaco," forestry begins during the 1880s in the eastern zone—along the strip bordering the Paraguay-Paraná riverine axis—and in following decades will extend toward the west. One of the factors that encouraged this activity in the western Chaco was the decision of the British railway companies at the beginning of the century to abandon the use of iron railway ties for ones made of *quebracho.*[6] That, together with the demand for posts for the wire fencing on the properties that the agro-export development of the Pampa required, stimulated forestry in the region.[7] Its position as the final point in the east-west growth line, both in terms of the number of trees as well as the variety of timber, explains why the intensification of forestry activity occurs in the *Llanos* while the Chaco's other western forestry zones—closer to the urban markets of the region—had already declined, something that occurs according to my calculations during the 1940s, an estimation supported by other scholars.[8] According to at least one author, in these years there occurs a displacement of the "forestry border" toward the interior basin of the *Llanos* of La Rioja and northern San Luis.

The *Llanos* of La Rioja

The province of La Rioja is situated in the Argentine northwest (see map I). Its economy is one of the poorest in the country and is comprised of three sectors, sharply demarcated by geography: the urban tertiary system (principally the provincial capital); the agriculture of the intermontane valleys (wine and olive production); and the pastoral economy and forestry economy of the *Llanos.*[9] La Rioja's *Llanos* cover approximately 50 percent of the province (4.6 million hectares) and form part of an internal plain bounded by mountain chains and extending west to the Andes. A region characterized by dryness, there are no rivers that carry water to the sea. Rainwater is quickly dissipated; the heat is intense and water rapidly evaporates. The annual 350 millimeters of rain that fall do not have a homogeneous distribution and make impossible the cultivation of crops without irrigation. The only agricultural activity of significance is livestock, though its importance at both the provincial and national level is small.[10]

Logging in the Gran Chaco's forests has led to widespread environmental destruction with especially severe consequences in the region under study.[11] The virtual elimination of tree covering has caused an increase in sunlight on the lower strata of flora (bushes and grasses), and thereby undercuts the forest's natural protective canopy during torrential rains, something that in turn has brought as a consequence the loss of nutrients and moisture in the soil. The destructive effects of logging on the *Llanos*'s ecosystems help explain the crisis in the region's livestock industry, a crisis that only a handful of big ranchers have managed to overcome; following the era of massive logging, these landowners managed to gain access to technology appropriate for an arid zone while those who practiced subsistence ranching either live in extreme poverty or have been forced to emigrate.

The Railroad and the Rise of Forestry in the *Llanos*

During the second half of the nineteenth century, as railroad lines were expanding almost exclusively in the Pampa region (the center of the country's agro-export economy based primarily on grains and meat), the economies of the northwest provinces continued to be tied to the markets of bordering countries (Salta and Jujuy with Bolivia, Mendoza and other Andean provinces with Chile). With the arrival of the *Ferrocarril Central Norte,* the Tucumán sugar-producing region, a region that had been experiencing a process of sustained expansion since the 1860s, became integrated into the urban markets of the Littoral.[12] Various branch lines would be added in the decades after 1870, thereby allowing the regions of the northwest to begin to send their products to the principal markets of the Pampa zone. This would serve to redefine the shape of the regional economies; some products would be given priority while others entered a notable decline. Some interregional ties would become important while others stagnated.

La Rioja, however, did not participate in this process of the formation of an internal Argentine market. Its most important economic ties continued to be with Chile. Since the beginning of the nineteenth century, there had existed a commercial route for cattle on the hoof to Chile, in which the *Llanos* zone of La Rioja fulfilled the role of raising the livestock, whereas in the intermontane valleys the cattle were fattened and then exported through the Copiapó and Jagüel passes with a final destination in the so-called Norte Chico region of Chile.[13] La Rioja would be the last province and last provincial capital to become connected to the national railway system. This finally takes place in 1891, the result of the interest of a British company ("Famatina Development Company") in mining the gold found in the Famatina Mountain, an undertaking that would fail years later.[14] Nevertheless, the railroad network did manage to make valuable a resource heretofore almost exclusively used for local consumption: forestry products. The wine industry of Mendoza and San Juan needed wood for trellises, vats, and barrels; the Pampa's agriculture for the posts for its wire fencing; the urban centers of the Littoral demanded firewood and charcoal for home use (heating, cooking, etc.) and for some small industries such as bakeries.

Extractive undertakings stimulated the expansion of the lines of the *Ferrocarril Central Norte* in La Rioja. In 1911, the second trunk line was opened, this one crossing the southern *Llanos,* linking western Córdoba from Serrezuela to the city of San Juan. Its construction was closely related to the emergence of the Cuyo wine industry. Between 1935 and 1938, the railway stations of the third line were opened, built between Milagro and Quines, the latter a railroad junction in the province of San Luis. The railroad's expansion here was due to the rise of the forestry industry in the *Llanos,* which increased beginning in 1939 thanks to the lack of fuel in the country caused by the Second World War.[15]

The "Obraje Cycle" in the Llanos

The so-called *obraje* cycle is the historical period in which the principal economic activity of the region is forestry, a period in which there can be demarcated moments of take-off, boom, and decline. One might mark as the beginning of the take-off with the arrival of railroad to the region (1891), since it is the railroad that will be the principal consumer of the firewood and that generates a sustained demand. However, quantitative evidence for this economic expansion has only been discovered beginning for the year 1901, the first year in which freight moved by train is registered in the publications of the *Dirección de Ferrocarriles Nacionales.* The decline, in turn, begins in the 1960s, as can be inferred by analyzing this same statistical information and the tonnage of wood registered by the IFONA (*Instituto Nacional Forestal Argentino*) for the entire province (1954–1980).

In analyzing the freight statistics of forestry products from the *Llanos* carried by rail between 1901 and 1980, there are two clear trends. The first, the increased volume between 1901 and 1940, the apogee of which can be considered the high point of the *Llanos*'s production, one that appears late compared to other major forestry regions such as those of the provinces of Chaco and Santiago del Estero. The other trend is the decline of production between 1940 and 1980. Since the statistical information is incomplete for this period and, moreover, is contemporary with the period when the railroad is replaced as the major freight carrier by motor transport, I have complemented the railroad statistics with those provided by the IFONA at the

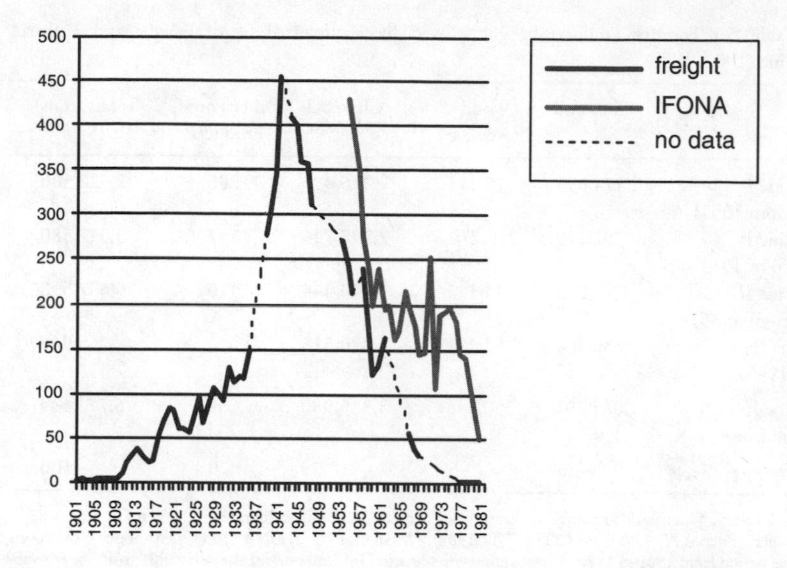

Graph 5.1 Forestry Extraction in the *Llanos* (1901–1980) (Measured by thousands of tons)

provincial level and further broken them down to the *Llanos* region to corroborate the downward trend, albeit a trend postponed to the following decade according to IFONA's statistics. (see graph 5.1)

The second trend is that the territorial distribution of production was uneven for the three railway lines. Line II, which links Córdoba with San Juan, is the line that transports the greatest volume of lumber, more than 3 million tons for the entire period. It is also the line that shows an average annual volume greater than the other two lines—65,051 tons annually (one and a half times greater the average of Line I running from La Rioja to Castro Barros, and almost three times the amount of Line III running from Quines to Milagro). By the same token, it occupies first place in the average annual kilometers traveled: 412 annual kilometers versus 246 for Line III and 183 for Line I. With regard to the kind of freight carried, wood for fuel was the predominant use and charcoal unquestionably occupies first place, comprising some 79 percent of the total; and if we add firewood it eventually comprises 95 percent of the production transported by rail (see Table 5.1).

In summary, we are witnessing an economic cycle with clear trends: a high point in production is reached around 1940, a production that was devoted primarily to produce wood for fuel in the form of charcoal and firewood,

Table 5.1 Forestry Products Freight Moved by Railroad: Type of Freight and Railroad Line (1901–1980)

	Firewood	Wood (Posts, etc.)	Charcoal	Other Forest Products	Totals (tons)
Line I (from 1901)	174,695	77,111	2,276,496	208	2,528,500
Line II (from 1911)	792,279	226,270	2,298,744	287	3,317,580
Line III (from 1935)	81,873	35,393	540,444	110	657,598
IFONA (1953)	36,364	12,084	220,518	0	268,966
Totals (in tons)	1,085,201	350,858	5,335,980	605	6,772,644
Percent	16	5	79	0	100

Sources: C. Natenzon, "El manejo de los recursos naturales renovables durante un siglo en Los Llanos de La Rioja", Buenos Aires, *Informe* CONICET (1987), "Memorias" de IFONA (1953). The above table covers the period from 1901 to 1980. There is complete statistical information for the years 1901 to 1935. For subsequent years, statistical information exists for 61 years, whereas there are 19 without any information. To this has been added the data taken from the "Memorias" of the IFONA which goes up to the year 1953, in which statistical information is broken down by rubric but not by railroad stations.

whose epicenter was the southeastern region of the *Llanos*. At the same time, important changes will be germinating as a result of the forestry activities, as becomes clear if we view these changes from the perspective of the principal protagonists of the region: the *obrajera* commercial firms, the forestry workers, the hide traders, and small rural producers.

The establishment and growth of commercial enterprises devoted to forestry dates from the first decades of the century and continued to be important in the 1940s, 1950s, and, to a lesser extent, the 1960s. We are dealing here with firms characterized by product diversity, generally operating at the retail level, which in these years went by the name of "general goods stores," and which included the importation via the railroad and sale of grains, cloth, sundry ironware goods, and purchase of hides in the rural areas (in the *barraca de cuero* or "leather stalls") and other goods of a household origin, as well as being involved in the forestry business, a business encompassing everything from the sale of forestry products to their extraction and gathering. In some cases, these commercial establishments also included the purchase/sale of livestock, principally cattle.[16]

La Rioja's commercial license census for the year 1924 shows the early existence of this group of merchants tied to the forestry business.[17] Out of a list of 54 commercial firms involved in the forestry business that appear in a

1943 report,[18] 30 already were active in 1924, at the time involved in petty trade, principally carrying lines of goods in textiles and foodstuffs.[19] By midcentury, these businesses had not abandoned the lines that constituted a "general goods" establishment, nor had they abandoned the "leather stall" marketing, while at the same time they had tended to increase the importance and scale of their involvement in the extraction, gathering, and sale of forestry products.

These commercial houses comprised various shops and stores, generally in the hands of one family, often of Arab immigrant origin.[20] Such businesses usually consisted of a principal store—the major capital investment of the enterprise—located near the railway station, and of other smaller establishments in rural hamlets. Both in the petty rural commerce and in the business transacted in the stores and the *pulperías*[21] at the railroad stations, bartering was a common form of exchange, a fact that reveals the constraints that existed for the development of a market economy in the region, given the importance of household production in the local economy. To take just one example among the many available, a traveling salesman who covered the countryside would exchange manufactured goods for hides and cattle, without an exchange of money. By way of anecdotal illustration, one regional observer of an itinerant merchant's travels commented to the author:

> He (the merchant) needed to return to Chepes because that trip had exhausted his supply of goods, having exchanged them for hides that had been acquired in the *estancias* en Nepes and also in Cortadera. He returned with cheese, "orejones" and even sheep hides and flasks of cider.[22]

Wheat, seeds, and tools could either be paid for with money or exchanged for cattle, hides, artisan goods, etc., if specie was short. The only economic activities which required money, or lacking that, letters of exchange, were those regarding trade outside the region, the contracts signed between the commercial houses and Cuyo wineries or the *Ferrocarril Central Norte,* or debts to outside creditors. According to a report of 1943:

> The majority of the "obrajeros" were advanced goods and financed by the consignees of the forestry products in the Federal Capital, to whom are presently owed great sums of money—the result of the stagnation in the supply of fuel—with their investments in the Llanos estimated to be some 4 million pesos … it can be assumed that the "obrajeros" activities are dependent on the limited amount of credit granted by the consignees.[23]

The members of these family businesses who ran the smaller branches in the countryside were generally authorized to act as agents by the eldest son or father—the main boss found in the head office in the urban centers or at the railway stations.[24] The records of the commercial registry show that these various branches were managed as a single business with regard to banking and financial operations, although the precise composition of the enterprise fluctuated, with the departure of a son or brother from the original enterprise, for example, or through the association of some member of the family with another local commercial establishment.[25]

An important sector in which the merchants worked was the hide trade (*barraca de cuero*). This was a business devoted to the purchase from local producers of hides, mainly cattle hides, sold either directly or through the branch stores and bought either from village markets or *puestos* found in the rural areas.[26] In turn, the merchant's authorized agent could either be the storeowner or might at times be a local producer himself, who could have either a fixed residence or might lead a peripatetic existence traveling the countryside to purchase hides. When the activities of the traveling merchant combined with that of the cattleman and drover (transporting cattle on the hoof), the latter bought cattle to provide them to urban suppliers. Whether the authorized agent was himself a rural producer or someone doing business with the rural producer, rural producers were coming into direct contact with merchants. This form of exchange caused the producers to become dependent on the merchants, both with respect to the goods they would receive as payment and on the cash advances they would be granted in anticipation of the price of the hides.[27] As a result, there occurred a capitalization of the "fruits of the countryside." The products of a household economy, in effect small cattle ranches, became important commercially for the urban economy taking shape in the railroad stations. As a corollary, these household economies, which until that point had been to a considerable degree isolated, now were becoming linked to the urban centers.

The terms of incorporation of one of the commercial enterprises in 1946 declared a capital worth of 100,000 pesos. In 1949, this same limited liability company had a capital value of 250,000 pesos, of which 125,000 pesos was contributed by the son and heir of the founder of the business and the rest by another partner.[28] Generally, the terms of capitalization were tied to the transformation from what, in effect, were already limited liability companies to a strategy of creating partnerships among different family commercial businesses.[29] As is revealed in the notarization papers that these partnerships sign, the capitalization is expressed in money, lands, forestry products gathered, tools and equipment (a truck, for example, would be noteworthy because of its relative value), and *haciendas* (livestock). One of the mechanisms that strengthened the process of accumulation was this system of consignment, which involved not only local merchants, traveling salesmen, cattlemen, and hide traders, but also the consignees in the federal capital. Through a network of various forms of exchange, the *sociedades obrajeras* managed to generate and sustain a process of accumulation, which led to the increasing influence of commercial capital within a regional society in which capitalist relations did not predominate.

Access to the Forestry Resource

How did the *obrajero* business gain access to the forest? Although the *obrajero* sector bought land, the most common way the *obrajero* became a lumberman was by renting land, principally those held in usufruct by peasants known as *derechosos* (the holders of rights to use the land's resources). The *derechosos,* although they were not the owners of the land, paid a territorial tax to the state as a way of validating their usufruct rights to the land.[30] The renting of forest lands—called in the colloquial language *derechos de monte*—was an arrangement arrived at through verbal agreement. The *obrajero* renter acquired the right to exploit the forest, without specifying which species of tree would be cut down, over an area with loose boundaries in return for a cash payment. Given that what was being used was the land's forestry resources, these renter's agreements were arrived at every time there occurred an advance of the extractive frontier. The owners of the *obrajes* were continuously obtaining new *derechos de monte* as the boundary expanded, and the expulsion of the *derechosos* was not a prerequisite to getting control of the forestry resources.

Organization of the Labor Process in the *Obraje*

Once access was acquired to the forest resources, there began the extractive process. The priority was to extract the "green wood" of which the *quebracho blanco* tree was the principal source, but which was also found in the carob tree (*algarrobo*). Both these species of trees possessed, on the one hand, a high caloric value, while at the same time met the required dimensions of the railways (for stakes, etc.). The unused part of the felled *quebracho* tree (the smaller branches and tree trunks), together with the byproducts (tar, broom, etc.) of other species were used in the carbonization process carried out in the forest itself.[31] With the *obraje* there emerges for the first time in the region's economic history, extractive processes of primary materials based on wage

Table 5.2 La Rioja: Composition of the Labor Force in the Forestry Sector (1947)

	Owner	Self-Employed	*"Ayuda Familiar"*	Employee	Worker	Apprentice	Total
Number	121	75	84	52	1,297	612	2,241
Percent	5.4	3.3	3.7	2.3	57.8	27.3	100

Source: Unpublished tables of the 1947 National Census, p. 221.
*The category "ayuda familiar" is not explained in the census, but it refers to unpaid household work, that is, a particular kind of self-employment in the individual's home itself.

labor, something that can be seen in the following table in which the category of "worker" together with "apprentice" comprise almost 85 percent of the population occupied in this economic sector (see table 5.2).

In the *Llanos,* "great obrajes" were established in labor camps containing between 100 and 200 wage-earning workers who labored in work gangs of between 3 and 9 men for each specialization. These specializations reflected the division of labor in the process of carbonization of the *quebracho.* The lumberjacks or *hacheros* cut down the trees. The *trocador* cut the felled trees into the appropriate sizes and the *rodeador* carried the wood from the place the tree had fallen. The *apilador* then arranged the logs of wood in mounds, while the *aguatero* distributed the daily rations of water to the workers. The *labrador,* in turn, prepared the logs, removing the bark and giving them the desired form. Finally, the *armador del horno* prepared the piles of wood to be used for charcoal and the *quemadores* had the task of overseeing the oven's combustion, a task considered to be particularly important for ensuring the maximum efficiency.[32] Between the stage of the carbonization of the wood and the transporting of the charcoal to the railway station, the work was the responsibility of the *tapadores,* who had to cover the oven with weeds and broom, according to the instructions of the *quemador;* the *sacadores* then removed the charcoal from the oven and the *tabicadores* carried the charcoal to whatever means of transportation being used.[33] At first, mule-driven carts were employed. Then, with the advent of motor transport in the province between 1935 and 1938, trucks began to replace carts driven by animal power for purposes of transporting the wood between the *obraje* and the railway station.[34] This means of transportation assumed even more importance after 1945, when army surplus vehicles entered the country, and then with the development of a national automobile industry in the 1950s (Industrias Kaiser Argentina and Industrias Mecánicas del Estado), the building of paved roads and the deteriorating service offered by the railroads, which encouraged the use of trucks between the gathering point and the market.

The forestry worker was brought in and employed (*conchabado*) by the *obraje* contractor and subsequently was also recruited among the resident population. The contractor directed the technical part of the productive process and hired work gangs; it was he who was responsible for hiring and paying the work force.[35] As long as the railroad was the preferred means of transport, the contractor obtained railway cars for the loading of the wood; but once trucks came into widespread use, he purchased these instead. The contractor operated in various ways. He could either work for the merchant, work for a bigger *obraje* contractor as a kind of subcontractor, or work for himself, with direct agreements with the railroad. In this latter case, the line between contractor and merchant became blurred.

Company Stores, Small Producer Debts, and Proletarianization in the *Obrajes*

The *obraje* company store was established in the countryside with the object of supplying goods to the work camps that developed around the extractive process. It preserved certain characteristics typical of Argentine rural commerce early in the twentieth century: it was closely linked to the commercial establishments at the railroad stations through a system of consignment or advanced credit on merchandise; it was involved in the hide trade; and it sold manufactured goods and bought goods of household origin. In terms of the latter, the population of subsistence rural producers in the surrounding area began to sell its small agricultural surplus to *obraje* workers and to the company storeowners, obtaining cash in return. The *obraje,* in turn, utilized its trucks to sell household-produced goods in the urban centers. As a result, certain household-produced goods entered the commercial circuit. Other goods formerly produced in the household were replaced by ones the company store brought from the outside, something that, in some cases, reoriented the household economy. Rural producers began in this way to become integrated into the urban setting and its commercial networks. The *obraje* was the center that mediated these commercial transactions. Commerce reached such a level of activity that the state began to feel the need to regulate it:

> It is urged ... that there be established a sentry station in Mascasin on the Camino Nacional to control the trucks that leave and enter with contraband, carrying forest products, goats, cheese to avoid paying taxes while at the same time bringing in goods to sell ... there is produced in the Chepes region approximately 100,000 kilos of cheese annually, both cow and goat cheese, the greater part of which is transported, in mockery of the border controls, as contraband by making use of trucks and the good road.[36]

A distinctive characteristic of the company store was the decisive role it played in the process of drawing the local population of small rural producers into the market, and in the control and subjugation of the labor force that was semi- or even completely proletarianized. For that purpose, two methods were employed: the voucher and the credit account. The voucher replaced money wages to the worker, guaranteeing in a coercive way purchases in these establishments, at prices stipulated by the merchants, which, according to reports, were significantly above market prices.[37]

> I have been informed that [the workers] are paid with vouchers or buy on credit and that they are obliged to make their purchases in the stores indicated by the *obrajeros;* burdened in this way with the price of the merchandise, the workers entire earnings return to the hands of the merchant or contractor.... [The *obrajeros*] ... have cantinas and supply the peons themselves, it is presumed that in the credit notebooks each worker is overcharged, for the simple and clear reason that when the personnel is paid, these receive nothing, and those who complain are fired.[38]

The peon became indebted to the company store, an indebtedness that is closely tied to his subjugation since he had no other way to pay off what he owed other than his own labor. There was thus created a true debt peonage.[39] This indebtedness led to the loss of land for thousands of small producers and *derechosos.* I have documented this process in the property assessment declarations of the *obrajeras,* as well as in the property registries where the titles of the current property holdings are to be found. By way of example, the history of the properties of the *obrajero* "San Jorge Comercial S.R.L." is instructive. In this particular case, in the territory corresponding to the land holdings, the property rights of those who held purchase agreements (*ausentistas*), were superimposed on those of the holders of usufruct rights (*derechos y acciones*). The process of land alienation worked as follows. First, the *obrajera*o rented land from the local *derechosos.* Once the forest was cut down, he bought various parcels and cleared title to them. Meanwhile, he bought the *derechos y acciones* with vouchers from the company store. In this way, the superimposition of categories in the land registry was abrogated, and with it the communal rights brandished by the peasants who ceased to be *derechosos* and were converted into "intruders." It is noteworthy that the sales transactions do not make reference to a total communal property but to small parcels found within this property, a fact that is meant to demonstrate the illegitimacy of the communal usufruct rights over the land.

Not only indebtedness served to "tie" peons to certain landowners. The company stores enjoyed a monopolistic position in commerce and the merchants also held a preponderant political power in the local setting. They competed with one another for the exercise of that autocratic style of rule over different urban and rural parts of the region. Soon, specific territories would be regarded as the preserve of specific business enterprises. Things reached the point that "town keys," symbolic tokens given to honored visitors but in reality symbols of the power acquired, were granted to the local economic elite. The town keys were the material symbols of informal verbal agreements reached between the *obrajeros* in the distribution of households in which they held a monopoly of commercial activity. This economic function was, in turn, associated with the exercise of local political power (in the police, municipal, or provincial governments). In these hamlets, the *obrajeros* regarded themselves as the "masters" of the peons who resided there, and from whom the *obrajero* recruited his labor force. In this context, disputes occurred for the control of entire peon groups. Even if there formally existed an absolute freedom to break the labor services tie, landowners considered rights to certain workers as their own. In this respect, a document written by a functionary of the federal government sent to "intervene" the province after the 1943 revolution reveals a great deal about one such individual and, it can be surmised, the behavior of these businessmen-*obrajeros*. This particular individual was the owner of most of the houses and adobe *ranchos* in Punta de Los Llanos. Availing himself to the fact that he was also the town chief of police, he took over the lands of some indebted *derechosos*, going so far as to appropriate and sell a cart full of corn sent by the federal government for the local poor in a relief mission. He argued that he held the "town key," something that obliged the peons of the region to work in his *obrajes* if the conditions themselves did not compel them to do so.[40]

Coerced Labor and Migration

The monetization of their incomes, indebtedness, and the loss of access to land were the end result of the *obraje* for a large part of the local population. Through the implementation of measures of a coercive nature, *obrajeros* attempted to facilitate the supply of a labor force to the extractive sector, converting the *derechoso* into a forestry worker. However, there did exist a factor that conspired against the availability of an ample labor supply: migration. The *Llanos* of La Rioja sent some 16,101 migrants out of the region between the two national censuses in 1914 and 1947, and 16,877 more between 1947 and 1960. This meant that the population's exodus intensified in relation to the size of the population itself, since the balances of net emigration increased from 13.52 percent in 1914–1947 to 30.1 percent in 1947–1960.[41] Some of these migrants settled in the expanding urban centers in the Littoral provinces, while others become involved in seasonal migratory flows. It has been estimated that some 6,400 day laborers from La Rioja left in search of work during the summer months (December to March) to "raise" the harvest on the Pampa (wheat and corn) between the years 1895–1914.[42] At the same time, the *Llanos* provided the "swallows" (migrant workers) for the Cuyo's wine harvest (January–March), and also for that of La Rioja's own intermontane valleys with their grape and olive production. In those same years, so-called "locusters" also were sent to the Pampa to combat that pest, forming "... a kind of levy of strong and able bodies" who were paid 80 pesos and their travel expenses. Moreover, contractors organized important contingents of workers for the Tucumán and Salta-Jujuy sugar harvests "... attracted by the novelty of seeing new lands and by the 50 pesos they were advanced."[43]

Although it has been noted that the work performed in these various regions required a nonskilled, seasonal labor force, there also existed some highly coercive characteristics in these economies as well, such as the use of wage advances for purposes of creating indebtedness and dependence, the system of vouchers and credit in the company stores, etc.[44] However, the information uncovered for select years and certain sectors indicates that it was predominantly wage levels and monetary incentives of various kinds that the surrounding regions offered, with their economies engaged in a vigorous expansion, which encouraged this tendency for the labor force to migrate. Early in the century, in the Salta-Jujuy *ingenios,* for example, harvest workers were making as much as 6 pesos a day (with lunch provided by the landowner) and in Tucumán up to 8 pesos a day (with lodging and food included), well above what they earned in the *obrajes.* In 1943, a *peón de cosecha* in Cuyo's wine harvest could make between 5.75 and 6.25 pesos a day, whereas his counterpart in the *obraje* earned between 3.50 and 4 pesos a day.[45]

The possibility of migrating to regions with higher wages, as well as the greater access to land outside the region, conspired against the availability of a local labor force, at least on terms favorable to the extractive industry's owners as regards work conditions and salaries. In this respect, one can even better appreciate the crucial role played by social practices of a coercive nature to attract and tie labor to the *obraje.*

After the *Obraje*

Due to changes in the country's energy needs and the exhaustion of the forestry resources, the industry declined and livestock replaced it as the principal economic activity of the region, though it was a livestock industry with scant significance at the national level. According to figures from the 1983 national agricultural census, the *Llanos*'s livestock constituted only 0.03 percent of the national roundup for that year, cattle moreover that were of an inferior mixed stock. The productivity per hectare and per inhabitant was also low in relation to other arid and semi-arid regions in the country.[46] The *obrajero* commerce, which during the extractive forestry cycle had accumulated lands and a large quantity of liquid capital, tended subsequently to invest in cattle raising. As a result of the appropriation of land, a process of land concentration had occurred in the hands of these same people who—in subsequent decades and supported by state subsidies—propelled certain changes of regional significance. Specifically, there occurred the adoption of technology that allowed landowners to counteract the arid environment, a characteristic of the region that had become even more pronounced with the process of desiccation resulting from deforestation.[47] Large tracts of land were fenced in and legal titles established. Agronomists specializing in cattle breeding and sanitary control were hired. Landowners experimented with artificial pastures and new bovine stock was introduced. Although technologies suitable to the arid environment were utilized, the general tendency was the occupation by a few estates of vast tracts of land, and extensive rather than intensive utilization of the soil, something that has led to extremely low labor requirements per hectare.[48]

The *obraje* had played a fundamental role in this process of proletarianizing the local population. The *obraje*'s company stores played a crucial role in the process of monetization of exchange in a society that heretofore had enjoyed a high degree of self-sufficiency. Similarly, the railroad had contributed to increase the mobility of the labor force, since it had put regions with different wage levels in contact with one another. Combined, the mechanisms that the dominant economic interests had employed for the attraction and subjugation of the labor force implied not only a loss of access to land (through debt) of those incorporated as wage-earning workers in the *obraje* but also a destruction of the self-subsistence practices and the economic stability of broad sectors of the local population. As a final consequence, it unleashed a process of temporary and in some cases permanent migration to other regional economies experiencing economic expansion, leading to a chronic or seasonal scarcity of labor. Natehzon summarized the historical process in the following words:

> the region's forests were extracted to be consumed elsewhere. Once it was, on the one hand, no longer necessary to resort to this source of energy and, on the other, the source—the woods—no longer contained elements of arboreal

worth, the social stratification of those who had lived from the industry became more pronounced. On the one hand, there were those who possessed the means to realize changes in methods of cattle production and, on the other, those who had to sell their lands, look for work in the government, or emigrate.[49]

Translation by James P. Brennan

Notes

1. Jorge Morello, "El Gran Chaco: el proceso de expansión de la frontera agrícola desde el punto de vista ecológico ambiental," *Expansión de la frontera agropecuaria y medio ambiente en América Latina* (Madrid: CIFCA, 1983), p. 343.
2. Jorge Morello, L.A. Sancholuz, and C.A. Blanco, "Estudio macro-ecológico de los Llanos de La Rioja," (Buenos Aires: IDIA, 1977), p. 7.
3. Tannin is a substance extracted from the *quebracho*—found to a greater degree in the *quebracho colorado* than in the *quebracho blanco*—which once had a variety of uses, among them tanning hides.
4. E. Bruniard, "El Gran Chaco," in *Revista Instituto de Geografía* (Resistencia: UNNE, 1975), pp. 44–45.
5. P. Denis, *La valorización del país: La República Argentina 1920* (Buenos Aires: Solar, 1987), p. 142.
6. J. Morello, p. 380. Another factor that increased the demand for wood is that in the Argentine interior the railroads used firewood for fuel, whereas in the Littoral they burned coal imported from Great Britain.
7. Between 1906 and 1915, a space no greater than 15,000 square kilometers in the Chaco zone of Santigao del Estero produced 14.5 million posts for the wire fencing of the Pampa. See Orestes Di Lullo, "El bosque sin leyenda; ensayo económico-social," (Santiago del Estero, 1937), p. 3.
8. F. Suárez, "La explotación del Bosque," in *El país de los argentinos* vol. 5 (Buenos Aires: CEAL, 1979), p. 675.
9. If one takes into account the 24 political jurisdictions in which Argentina is divided (23 provinces and the federal capital) La Rioja occupies the twenty-third position with regard to the value of its agricultural production and the twenty-first position in terms of the value of its per capita industrial production. See P.M. Aguilera, "La estructura económica de La Rioja en la década de 1970," in *La economía riojana: Realidad, política, estrategias* (Córdoba: Editorial de la Municipalidad, 1988), p. 141.
10. The *Llanos*'s livestock industry represented only 2 percent of the gross provincial product as calculated by the "Secretaría de Estado de Agricultura y Ganadería de La Rioja" in its "Informe sobre el sector agropecuario provincial" (1989), p. 2. The region's annual cattle roundup in 1988 was about 200,000 head, that is 0.02 percent of the national total.
11. According to Jorge Morello, logging has done such damage to the ecosystem of the Gran Chaco that it has now been irrevocably altered without possibilities of returning to its original state, *Expansión de la frontera*, p. 356.

12. Raúl Scalabrini Ortiz, *Historia de los Ferrocarriles Argentinos* (Buenos Aires: Plus Ultra, 1974), pp. 199–238.
13. Gabriela Olivera, "Transformaciones de la economía familiar en el contexto histórico de este siglo" (mimeo) (Córdoba, 1989), pp. 175–202.
14. E. Bocco de Brizuela, "Aportes para una historia del Ferrocarril en La Rioja, 1875–1986," (mimeo) (La Rioja, 1987), p. 23.
15. E. Bocco de Brizuela, "Aportes para una historia," p. 23.
16. Archive of the Tribunales de La Rioja (henceforth ATLR), "Libros de Registro. Público de Comercio de La Rioja," Years: 1940–1946, 1946–1950, 1956–1960, 1960–1966.
17. Archive Histórico de La Rioja (henceforth AHLR) "Padrón General de Patentes del año 1924." The marketing of forestry products does not figure in this census and that is due to the fact that this activity is taxed separately. According to a 1916 provincial law, forestry products are taxed according to quality and weight.
18. AHLR, "Informe de la Intervención Federal en La Rioja sobre el sector forestal," Año 1943, pp. 3–6.
19. In 1924, these firms' volumes of capital were small even by the standards of the province. In only two cases did these commercial establishments have a capitalization that can be considered important by provincial standards (between 8,000 and 10,000 pesos).
20. 85 percent of the total number of the owners of such businesses were "Arab" (Syrians or Lebanese) immigrants or sons of immigrants.
21. A *pulpería* is a retail store that includes the sale of alcoholic beverages (by the cup).
22. *Orejones* are dried peaches, prunes, or pears.
23. AHLR, "Informe de la Intervención Federal en La Rioja sobre el sector forestal," Año 1943, pp. 3–6. *Obrajeros* were the owners of commercial establishments that, though of a diverse nature, were especially engaged in the collection, extraction, and commercialization of forestry products.
24. Very few of these authorization agreements were notarized, generally being agreed to verbally. The explanation for this is that they were usually established between members of the same family business and were therefore considered the internal affairs of the company and the family. The few written contracts that were preserved are precisely those formalized between businessmen who were not family members. For example, the establishment of a country "general goods" store in Ulapes under the management of Abraham Farias, authorized by Salomón Sinffe with its head office in the Tello railway station of the *Ferrocarril Central Norte*. This latter businessman pledged to advance goods but did not establish a cash payment as a condition. Goods were either paid for up front or on credit. Prices were established according to the instructions given by Sinffe, with profits divided 65 percent for Farias and 35 percent for Sinffe. ATLR, "Escribanía de Registro de San Martín," Año 1935, La Rioja.
25. ATLR, "Libros de Registro Público de Comercio de La Rioja."
26. *Puestos* are small rural hamlets that also serve as stopping and watering places for cattle.

27. Research conducted on this kind of business arrangement indicates that prices for the goods supplied by the local producers were generally established above their real value, whereas the amount paid the latter for the hides were under their market value. "This created an enormous profit margin that, in precapitalist eonomies, is typical of the monopolist position enjoyed by commercial capital." J. Chiaramonte, "Mercado de mercancías, mercado monetario y mercado de capitales en el Litoral Argentino de la primera mitad del siglo XIX: el caso de la provincia de Corrientes," in *Anuario de la Universidad de Rosario*, 12, (Rosario, 1986–97), p. 102.
28. ATLR, "Libros de Registro Público de Comercio de La Rioja," Tomo 1940–46, Expediente no. 18695.
29. The six principal cases of partnerships of commercial enterprises that evolve in this manner are: the partnership of V. Seguí and J. Abelín; the establishment of the partnership of Flores Euclides and Otros (S.R.L.); González y Ayan (S.R.L.); the expansion of Taboada Hmnos.; Los Llanos Forestal, Agrícola y Ganadera S.R.L.; and the transformation of the Chepes company into the Floresta Chepes S.R.L.
30. The most common form of tenancy was occupation, with rights to common usufruct of the land without title, called locally *derechos y acciones*, a term taken from the colonial land grants (*mercedes*).
31. Museo Ferroviario Nacional, "Informe del Ingeniero Principal de la Dirección de Ferrocarriles," Año 1939, p. 146.
32. Interviews with E. Campos and P. Carbel, Chamical, Los Llanos de La Rioja, December, 1989.
33. Interviews with E. Campos and P. Carbel, Chamical, Los Llanos de La Rioja, December, 1989.
34. The replacement of the mule-driven cart by trucks increased the productivity of this industry. In every trip to the railway station, carts pulled by 4 or 6 mules could transport between 400 kg and 2 tons of forestry products, with a frequency of only once a day. Trucks could transport from 3 to 3.5 tons, with a greater daily frequency. With this the daily load rose to between 9 and 12 tons a day.
35. Payment could either be in the form of a daily wage or piecework.
36. AHLR, "Informe de la Intervención Federal . . . ," pp. 29–30. The "border controls" to which the quote refers were those of surrounding provinces.
37. It is worth clarifying that this practice was found among other sectors and in other regions, for example among the Tucumán sugarworkers, those who labored in Cuyo's vineyards and in salt collection in the central part of the country. These abuses prompted a Senate bill, which ultimately was not passed but which stated "it is obligatory to maintain prices at their market value" and that "any practice is unlawful that obliges services as payment for goods made available though credits in company stores." In J. González Iramain, "La concordancia riojana en el Congreso" (Buenos Aires, 1942), pp. 57–58.
38. AHLR, "Informe de la Intervención Federal . . . ," pp. 29–30.

39. On the subject of debt peonage in Latin America, see, for example, A. Guerrero, "Ensayo sobre la acumulación originaria en Ecuador: haciendas, cascaoteros, banqueros, exportadores y comerciantes en Guayaquil, 1890–1910," in *Orígenes y desarrollo de la burguesía en América Latina, 1700–1955*, ed. Enrique Florescano Mexico: Nueva Imagen, 1985; Ian Rutledge, "Cambio agrario e Integración: El desarrollo del capitalismo en Jujuy, 1550–1960," (Buenos Aires: Proyecto ECIRA-CICSO, 1987) (as well as Marcelo Lagos's criticism of Rutledge's thesis in this volume); and Daniel Campi, "Captación y retención de la mano de obra por endeudamiento. El caso de Tucumán en la segunda mitad del siglo XX," in *Estudios sobre la historia de la industria azucarera argentina*, ed. Daniel Campi (Tucumán: Universidad Nacional de Tucumán, 1991).
40. AHLR, "Informe de la Intervención Federal . . . ," p. 42.
41. I have calculated the intercensal migratory balances based on the information contained in the population censuses of the *Secretaría de Salud Pública de La Rioja.*
42. A. Lattes and Z. Reccini de Lates, "Estudio de las migraciones en la Argentina, basados en datos censales, 1869–1960," (Buenos Aires: Instituto Di Tella, 1969), pp. 128–129.
43. M. Bravo Tedin, "La Rioja en el Centenario," in *Historia de La Rioja* (La Rioja: Dirección de Imprentas, 1985), p. 128.
44. Daniel Campi, "Captación y retención . . . ," pp. 131–133.
45. These figures come from the following sources: Mininterio de Agricultura y Ganadería de la Nación, "Anuario Estadística de la República Argentina. 1900–1906," (Buenos Aires, 1907), p. 1775; J. Bialet Masse, *El estado de las clases obreras argentinas a comienzos del siglo*, vols. 1 and 2 (Buenos Aires: Centro Editor de América Latina, 1985); AHRL, "Informe de la Intervención . . . ," p. 22; *La Rioja,* 25 February 1943.
46. F. Forni and R. Benencia, "Estructura agraria, sistemas productivos, mercados laborales y dinámica poblacional en las regiones áridas y semiáridas de la Argentina," Buenos Aires: CEIL, 1984, pp. 42–48.
47. F. Menvielle, "Desertificación en zonas semiáridas argentinas. Identificación de indicadores," (mimeo) (Buenos Aires: 1987).
48. This extensive livestock economy requires, according to my calculations, a single permanent wage-earning worker per every 5,000 head of cattle. The demand for temporary labor varies according to the climatic season and the technological level of the enterprise.
49. Claudia Natenzon, *Agua, recurrencia social y organización territorial en los Llanos de La Rioja* (Buenos Aires: Instituto de Geografía de la Universidad de Buenos Aires, 1990), p. 230.

CHAPTER SIX

Catholicism, Culture, and Political Allegiance: Córdoba, 1943–1955

Jane Walter

There was a time when many Catholics in the city and province of Córdoba were Peronist, but that was not to last. On November 10, 1954 President Juan Perón publicly announced that clergymen were meddling in the nation's professional organizations. He went on to explicitly accuse Córdoba's Archbishop Fermín Lafitte and seven of his priests of antigovernment activities. Two other bishops, more priests from other provinces, and the Church's principal lay organization, Catholic Action, were also named in his charges. In addition, Perón professed incomprehension at the emergence of specifically Catholic professional groups and scoffed at the electoral chances of any nascent Christian Democratic Party.[1] Months of conflict followed between Church and state, culminating in September 1955 with a successful military-civilian revolt based in Córdoba against the Peronist government.

Diocesan priest Quinto Cargnelutti of Córdoba did not oppose Peronism in its early years, but he became a key actor in the struggle against Perón. Archbishop Lafitte brought Cargnelutti from a parish in the province's interior to the provincial capital in early 1954 in order to develop a youth organization that could compete with the Peronist *Unión de Estudiantes Secundarios.* Cargnelutti's efforts proved to be most successful, so much so that he figured prominently among the priests cited by Perón in his November 10, 1954 speech. Cargnelutti went on to participate in the final successful uprising in Córdoba against Perón.[2]

Yet other Catholics found it more difficult and painful to sort out their beliefs and allegiances. A Peronist and a practicing Catholic, Elvira Rodríguez Leonardi de Rosales had become a senator representing Córdoba in the national legislature in 1952. On June 23, 1954, in support of a bill reforming the government ministries, she declared:

> Nowadays the people of the Republic are witnessing the encouraging sight of a community that is aware of its rights and duties. . . . The spiritual unity that this presupposes, emerges from the profoundly Christian and humanist content of Perón's doctrine.[3]

Yet in December 1954 she suddenly resigned her Senate seat in opposition to a Peronist-sponsored bill legalizing divorce, a measure directly contrary to Church teachings. Rodríguez Leonardi was soon expelled from the Peronist Party and denied the right to return to her teaching position in Córdoba.[4]

Román Aníbal Ludueña was a municipal worker, an activist in the municipal workers' union in the city of Córdoba and also a member of the Church-sponsored *Juventud Obrera Católica.* He found little contradiction in his Peronist and Catholic beliefs until May 1955, when the national Congress was considering a Peronist proposal to seek constitutional separation of church and state. He recalls:

> . . . all the employees were required to sign their endorsement of this proposal, many people did not sign: ANIBAL, obeying his religious upbringing, refused to sign said proposal, only doing so under pressure from the comrades of (the union) and the politicians of the period.[5]

What was the relationship between religious identity and political allegiance for each of these people? In other words, how did the Catholic and Peronist perspectives become so intertwined that the Church-state conflict took some Cordoban Catholics by surprise even as others were at the very forefront of the struggle? And what made Córdoba such an early locus of Catholic anti-Peronist sentiment?

While there is a rich and diverse literature on Catholic-Peronist relations in Buenos Aires during 1943–55,[6] provincial perspectives on the subject have been sorely lacking. That gap is beginning to be filled for the case of Córdoba. César Tcach's excellent study of political parties in Córdoba shows the predominance of traditionalist forces—conservative Catholic elites, antiliberal nationalist Radicals, and Catholic activists—in Peronism in the province, especially given the lack of a large industrial working class there at the time. The strong Catholic presence led to a rupture in the Peronist leadership of Córdoba with the public emergence of tensions between Church and state.[7] Silvia Roitenburd has argued convincingly that the views of the nationalist and nondemocratic Cordoban Catholic elite concerning proper family relations as well as the system of values for the society and the educational system differed dramatically from those of Peronism. She finds that the Catholic elite had a complex relationship with Peronism that shifted from conditional support to open opposition.[8] Nevertheless, these works do not focus on how the Peronist message could resonate with Catholics, or what specifically in Cordoban Catholicism could have motivated the important early challenge it mounted against Peronism in 1954.

This chapter will argue that the Catholic Church and Perón each had integral views toward orienting society that overlapped but were not identical. Perón's project for the nation included protection of specific traditional Church interests, fulfillment of aspects of the Church's social doctrine, proclamation of the Christian basis of Peronist doctrine and practice, and exaltation of what were termed the "spiritual values" needed to construct the new Argentine and the new Argentina. On these bases, many people were able to combine their religious and political sentiments. Yet some Catholics began to have doubts about Peronism, especially as the arena of Catholic activity was diversifying and activists were mobilizing even as the state increasingly imposed the Peronist perspective as the source of values for the entire nation. The 1954 Church-state conflict in Córdoba grew from both activist Catholic resistance to a perceived imposition of Peronist cultural values, and a struggle within the Argentine Church, sanctioned by the archbishop of Córdoba, to define a more liberal Catholic social presence in contrast to past organizational styles. Finally, I will suggest that class background and organizational identity were the principal factors determining why some Catholics became anti-Peronist relatively early, while others remained Peronist until forced to choose one allegiance over the other.

This chapter is divided into four parts. The first section discusses how Catholic and Peronist cultural values seemed to share much in common for many, though not all Cordoban Catholics during Perón's rise to power 1943–46. The second section shows that significant gaps between Catholic and Peronist values were already evident during the period 1947–49, although these values still overlapped a great deal. The third section argues that the disparities between Catholic and Peronist values became more pronounced in Córdoba during 1950–53, as Church groups and the Peronist regime pursued different missions with respect to society. Nevertheless, the relationship between political allegiance and religious identity continued to be quite harmonious for some Catholic Peronists. Finally, the fourth section details how the public eruption of strife in 1954 between Catholics and

Peronists in Córdoba soon led to a full-blown conflict between Church and state over cultural values. For Cordoban Catholics still loyal to Peronism, the harmonious relationship between their religious and political identities suddenly was ripped asunder.

1943–46: Values Potentially in Agreement

"Integral Catholicism" was the overarching project dominant in Catholic thinking and practice worldwide from the latter part of the nineteenth century until the 1960s. This intransigent, Church-centered vision had originally emerged as a reaction against modernism and secularism. According to Fortunato Mallimaci, in the view of integral Catholicism the Christian faith constituted the source of absolute truth, everything of real value originated from it, and the Roman Catholic Church represented the supreme norm and only guarantor of the unity of Christian truth and true values. This vision of Catholicism did not accept being relegated to the sacristy and sought in many different ways to have a social presence, engendering a vast movement that utilized concepts like penetration, transformation, restoration, and infiltration of the society. The massive and expressly nonpolitical "Catholic Action" lay movement the bishops created in 1931 under their direct authority was perhaps the Church's major means of carrying out the goal of restoring Christian values to Argentina. However, the integral Catholic project was not tied to a particular Church person or group. Instead it more generally provided the goal of re-Christianizing society. Integral Catholicism thus included groups with an Integralist or nationalist perspective (seeking virtual identification between Church and state and thus imposition of Christian values from above), or those with a more social or liberal Catholic vision (wanting greater distance between Church and state and consensus-building within society), or those who simply had a purely integral perspective that only the Church held the Truth and Catholicism the answer to the world's woes.[9] Such views were generally held by middle- and upper-class Catholic activists, who were the leading proponents of a Catholic restoration in Argentina along with the clergy. These categories did not necessarily encompass the religious sentiments of people with more popular understandings of Catholicism and religion, or those whose Catholicism was much less central to their everyday identity.

As a corollary to its general mission, the Argentine Church was specifically interested in certain areas useful for retaining and promulgating its influence in society. These areas became the sites of Church political efforts.

In education, the Church sought Catholic religious instruction in the public schools and a minimum of government interference in its own schools. Activities with young people—educational and organizational—constituted a key arena for instilling Catholic values in Argentina's youth. With respect to protection of the family, the Church was against divorce, abortion, and equal legal status for children born in and out of wedlock, and wanted proper moral behavior to be displayed in films, print, dress, and the behavior of the citizenry. In its relations with the state, the Church sought retention of the general constitutional provision whereby the federal government would "sustain" or support Catholic worship, an end to government involvement in designating new bishops, the signing of a concordat with the Vatican, government financial assistance in running its institutions, and state acknowledgement of the Catholic identity of the nation with measures like official recognition of Catholic holy days as national holidays. The Church also looked askance at competing religions and ideologies such as Protestantism, liberalism, socialism, Nazism, and communism.

In addition, papal encyclicals had developed a Catholic social doctrine expressing support for construction of a more just social order. These teachings acknowledged that workers' wages should reflect the needs of the individuals receiving those wages, not just profit motives; that private property was acceptable, yet the needs of the wider community must also be taken into account; and, with the crisis of liberal capitalism in the early 1930s, that corporatist organization of society would enable workers, employers, and the state to collaborate in constructing a new, Christian social order. However, by the end of World War II, Pope Pius XII had begun to newly affirm liberal democratic political structures and social reform.[10]

Catholic sentiment was well-developed in the city and province of Córdoba, and would prove to be characteristic of many Cordoban Peronists. In reaction to the dominance of anticlerical liberalism beginning in the late nineteenth century, the Church promoted its presence in Cordoban society through activities like the establishment of Catholic schools (aided by the arrival and founding of male and female religious congregations), activism in lay Catholic groups (culminating in Catholic Action), and also participation by activist Catholics in secular groups and arenas. Participation by Catholics from the traditional university and landowning elites in secular organizations and spheres was especially important for Church interests because such people could achieve highly influential positions. There was a significant Catholic presence in both major parties from the start of party competition. Catholic influence was greater in the conservative *Partido Demócrata* than in the *Unión Cívica Radical,* whose major electoral base was the immigrant-settled south and east of the province. Progressive, nationalist, nonreligious Radical *caudillo* Amadeo Sabattini became governor in 1936 and was followed by another *sabattinista* in office. The powerful Sabattinista wing of Cordoban Radicalism opposed religious instruction in the public schools and spurned state promotion of Church interests.

The 1943 nationalist military coup against the conservative national government displaced the *UCR* locally and the new military rulers began instituting decidedly pro-Church measures such as placing crucifixes in official buildings and establishing religious instruction in the public schools. Prominent Catholics were also named to positions of influence. For example, Lisardo Novillo Saravia, president of the Diocesan Board of Catholic Action 1931–36, became *interventor* of the National University of Córdoba during 1943–45,[11] and Rafael Moyano López, a member of the Diocesan Board of Catholic Action in the early thirties, became president of the provincial '*Consejo de Educación*' during 1943–45.[12] Meanwhile Juan Perón, based in the national government's "Secretaría de Trabajo y Previsión Social," had begun to actively promote workers' social welfare and organization from a nationalist and Christian perspective rather than a Marxist one, as indicated in his November 1944 speech to the workers of Córdoba:

> Our revolution, which is underway, is taking shape behind things which are too sacred to surrender: Our banners are God, the Fatherland, and Social Justice. We follow God through the words of the Divine Teacher, making men love their fellow men as they would themselves; let all Argentines join forces behind that love, because only love is constructive. Struggle destroys values, men, and societies.[13]

Lila Caimari notes that in his public speeches at the time, Perón tended to emphasize the Christian source of his social policy and went on to explicitly relate his views to Catholic social doctrine during his electoral campaign for the presidency in late 1945. She argues that this Christian and Catholic emphasis was Perón's need to set his activist policies in a recognizable legitimizing context that would soothe the concerns of employers and the middle class, as well as bring him support from socially active Catholics in particular rather than the bishops.[14]

In the province of Córdoba, the Peronist presidential ticket (supported by the *UCR-Junta Renovadora*[15] and *Partido Laborista*) lost the presidential race to the opposing *Unión Democrática* alliance of parties, but won the race for governor over the *UCR* by 183 votes. According to electoral analysis, the key support for Peronism in the province came from lower- and middle-class urban sectors especially in the capital, tenant farmers in the south possibly motivated by a government decree in 1945 granting terms favorable to them, and conservative Democratic Party *caudillos* in the rural south and center of the province with influence over voters in their areas. Such conservatives were perhaps especially determined to prevent the *sabattinista* Radicals from obtaining victory in their zones and at the provincial level.[16] The clearly Catholic component of Cordoban Peronism came from the non-reformist and largely Catholic wing of the conservative Democratic Party that shifted into Peronism, the nationalists of the *UCR-JR,* some of whom were fiercely Catholic, and members of the clergy and activists such as in Catholic Action,[17] all of whom largely held nationalist or integral views. But Catholic sentiment was not a notable characteristic of the workers forming the Labor Party, allied with the *UCR-JR* in support of Peronism.

On a more personal level, the choice to vote for Perón was natural for Aníbal Ludueña, who was living in Buenos Aires at the time and working in a soft-drink factory. He had been impressed with Perón's policies from the start and readily joined the local trade union, going on to participate in the mass rally in support of Perón on October 17, 1945.[18] Elvira Rodríguez Leonardi de Rosales belonged to the Radical Party and did not actually join the Peronist Party until later, after going to Buenos Aires to seek assistance for her school from Eva Perón.[19] Erio Bonetto considered himself a popular nationalist; he was from a non-Sabattinista Radical family whose Catholicism was open and tolerant rather than tightly connected to the Church and sectarian. He recalls that, among other things, Peronism provided an opportunity for him and his friends to quickly advance in their political careers, which would not have been as possible in the more established parties.[20] Lídia Torrez de Blangetti, who was living with her husband in the eastern part of the province, recalls that they switched from the Radical Party to Peronism because the Estatuto del Peón established a minimum wage for rural workers like her husband.[21] During the emergence of Peronism, Francisco Compañy was working as the parish priest in the town of Ballasteros. Various influences motivated him to support Perón. The *estibadores* (freight-loaders), *peones,* and other workers in his rural parish helped him understand the importance of unions and the need for the government to construct housing for rural workers, thus diminishing the loss of population from the countryside. Resistance in the universities in 1945 to the *revolución popular* and the need to retain religious instruction in the public schools confirmed his entrance into Peronism. Compañy was also impressed by how Perón responded shortly before the election to questioning about how he would combat communism: get rid of the causes by granting laborers greater and more solid advantages than those promised by the communists.[22]

Yet the Cordoban Catholic elite in the capital divided on the issue of support for Perón. Lisardo Novillo Saravia and J. Maldonado Lara, the archbishop's legal representative, became Peronist.[23] However, the conservative Catholic newspaper *Los Principios,* also linked to the archbishop's office, did not. Viewing measures like state support for higher wages and increased benefits for workers in disputes with employers, its editors were concerned about a possibly excessive redistribution of resources. When the government's regional labor delegation in Córdoba put an end to Catholic union organizing on the basis that all spiritual content should be eliminated from the unions, *Los Principios* noted that this meant communists would be able to take over the labor movement. The newspaper believed that the *Secretaría de Trabajo y Previsión*'s support for the designation of a single union for each branch of industry was an infringement upon the right to associate and unionize freely. It also echoed a Cordoban employers' declaration against state intervention in the economy and in favor of free enterprise and the return of institutional normality. The editors similarly disliked how Perón kept blaming "the oligarchy" for the nation's problems, noting that his words instigated hate and class struggle more than they fulfilled the social doctrine of the Church. *Los Principios* came out against Perón in the presidential contest, despite the fact that the *Unión Democrática* opposition coalition included in its program the reestablishment of laic (nonreligious) instruction in the public schools. The editors also were favorably disposed toward the Democratic Party's gubernatorial candidate.[24]

Córdoba's incipient Christian Democrats, who had a more reformist or liberal vision of Catholicism than did *Los Principios,* came out solidly against military intervention in politics, seeking instead a true Christian democracy based on law and liberty.[25] They did not find Perón's candidacy for presidential office at all propitious for the fulfillment of their goals, and announced their support for the Union Democrática's candidate. The alternative—Perón—would mean totalitarianism.[26]

1947–49: Overlapping but Not Identical Values

In April 1947, the Peronist government fulfilled the much-desired Church goal of converting the 1943 decree instituting religious education in the public schools into law. The task to pass *enseñanza religiosa* had not been easy, due to the resistance of working-class Peronist deputies in the national Congress with leftist and anticlerical sentiments. The Permanent Commission of Bishops[27] visited President Perón to express Argentine Catholicism's satisfaction and gratitude for passage of the measure, giving him a document to mark the occasion that quoted a speech he had given to an international audience of Catholics in October 1946:

> ... Our education must return to give priority to those things which make men great ... may our schools develop, not only intelligent men, but good men and wise men. Men who love truth more than power, reason more than force, and, above all things, who have a love for God, faith in his actions and hope in the future; a hope that we all in our infinite insignificance place in Him, in his infinite greatness.[28]

Yet Perón's religious or spiritual vision went further than just the Catholic instruction for which he was being thanked. Perón responded to the bishops by launching into a speech about the spiritual and humanist sense of the movement inspiring his government, noting that it drew from elements of Christian social doctrine. He also spoke at length about the doctrine underpinning his movement, confirming the concepts of justice, individual freedom, and Christian charity that provided the essence of modern Argentine spiritual formation.[29]

There were at least four distinct yet interrelated threads of Peronist spiritual discourse. One involved visible and concrete protection of Church interests, such as through congressional endorsement of religious education in the public schools or government financial support for Catholic schools. Another thread tapped Catholic or Christian social doctrine as the basis of Peronist reforms to improve the material conditions of people's lives and thus achieve social justice.

Proclamation of the authentically Christian basis of Peronist doctrine and practice constituted the third thread of Peronist spiritual discourse. The difference here from the second thread was that here Perón explicitly put forth himself, rather than the bishops or clergy, as the authority defining the true essence of Christianity. This tone was evident in his private letter to the pope of March 28, 1947 notifying the pontiff of Argentina's new law ratifying religious instruction in the public schools. Perón went on to complain that a troubling minority of clergy and bishops did not support his efforts on behalf of workers' rights, thus wresting them from the grip of leftist ideologies. He then called for the assistance of Church figures with his work of social regeneration:

> The peace and well-being of a people ... are not achieved simply through economic improvements, nor in comfort, nor in a rise in the standard of living.

Along with material conquests it is necessary to form the spirit.... It is necessary to fashion men in the practice of evangelical virtues, principally Christian austerity.... I hope for the determined collaboration of the Distinguished Bishops and Priests of my country, who are not adhering to any political cause by teaching the social Christian doctrine, by devoting themselves selflessly to the instruction of the people, to the re-Christianization of the family, to the defense of the oppressed, of the weak, of the wage-earner.[30]

Perón went on to publicly explain his perspective on Christianity to the Church when he singled out Bishop Nicolás De Carlo of Resistencia to be awarded for his work with the poor in a ceremony on April 10, 1948. Quoting extensively from the New Testament and revered Catholic writers, Perón took it upon himself to distinguish between good and bad Christianity. Caimari has pointed out that such a presumptuous attitude toward representatives of the institutional Church was essentially anticlerical:[31]

Our religion is a religion of humility, of resignation, of the exaltation of spiritual values above material ones. It is the religion of the poor, of those who feel a hunger and thirst for justice, of the disinherited and only for those causes which the eminent Prelates well know... there has been a subversion of values and an acceptance of the casting out of the poor of the world in order for the rich and powerful to take control of the temple.... I am proud to have made the Ministry of Labor and Social Welfare, a place where all enter with the same rights and that if there are looks of sympathy and comfortable chairs, that they go to those wearing humble clothing, to those "descamisados" rich in faith, despite the hardships in their lives, and who have been mocked with sinister political intentions.[32]

From this perspective, Peronist social practice would be the key to widespread fulfillment of the true Christian message.

The fourth thread of Peronist spiritual discourse was exaltation of what were termed the spiritual values needed to construct the new Argentine and the new Argentina. One element of such values involved development of a '*conciencia social*' in the country, using a three-pronged strategy:

For the owners to humanize capital in order for the country to preserve its human resources so that these human resources work with their hearts and not just their muscles, that these be rewarded for their efforts in order for mental and physical weariness to be minimized. For the state to dignify work by dignifying the worker because those countries where to work is the sad fate of the unfortunate and not a virtue of one's people are destined to succumb. For the worker to elevate his social culture because in the elevation of the social culture of the workers is the true dignity of the nation.[33]

A related spiritual element involved education, or the training of "useful citizens … competent and virtuous … wise and prudent … reflective men and men of action … moral men who learn that in order to be free it is necessary to know how to be slaves of duty and the law … men who know that the road of life is conquered through sacrifice and through honor, never through lassitude nor by satisfying vice and passions."[34] In fact, the Peronist project of integral reforms for society was so vast that it would require "a new philosophical school charged with forming a new soul to replace the old one, where truth, tolerance, wisdom and prudence are the foundation and pillars which uphold this new conception, one we must teach to our descendants for the greatness and honor of our fatherland."[35]

Juan Perón's wife served to reinforce and extend Peronist spiritual discourse. As leader of the campaign to obtain female suffrage, Eva Perón pointed out that the Argentine woman's deeply Christian identity as mother, wife, or daughter would lead her to vote wisely. She went on to liken the mother's role in raising tomorrow's citizen to the role played by *enseñanza religiosa* in the school system:

> Think how for the formation of tomorrow's citizen that there is no better school than the one run by the mother who votes …. Just as religious education grants to education a moral law which justifies and supports it, so there is the woman's influence who, in every home, is shaping in the mind of her child, a political consciousness of order and respect for institutions.[36]

Besides displaying support for religious instruction in the public schools, Eva was also affirming the traditional model of womanhood, one which the Catholic Church advocated as a critical support for protection of the family.[37] Like Perón, Eva declared that the Peronist "Revolution" was based on the Christian concept of social justice; in fact, "General Perón … is the apostle of social justice …."[38]

Eva Perón was likewise certain about Peronism's special closeness to the divine being. Thus she advised the oligarchic classes that God would help Peronism in all that it might do to sustain "la causa del pueblo."[39] On another occasion she declared that God had sent Perón to work on behalf of the people:

> Today in a popular demonstration, I happened to overhear a dialogue between two workers from Tucumán. One said to the other, "I love Perón more than God because Perón gives me everything and God nothing." The poor man forgot that it was God who sent us Perón in order to bring well-being to the homes of the poor. We owe thanks to God who remembered this

> long-suffering and generous people, putting Perón here to lead them in order that he guide the fatherland's destiny and govern in behalf of all Argentine and not just for a hundred privileged families.[40]

By extension, Perón was doing God's work. Eva also asserted the value of her own efforts in the area of social welfare in contrast to the benevolent assistance practiced by elite women—often Catholic—in the past. Thus her foundation for social assistance carried out not charity but rather justice.[41]

With respect to spiritual values, Eva Perón manifested absolute dedication to the Peronist cause through total identification with Perón and '*el pueblo*,' and complete devotion to her work with the poor and with women. Thus on Christmas Eve 1947, Eva told of her desire to be symbolically with her *descamisados* wherever they might be to celebrate that special night in a new Argentina where justice and happiness now reigned. As someone from their same background, she wished for no other course of action as the president's wife than "this, the sweetest, toughest, most marvelous one of all: to be with my heart beating, side by side with all of you. To be with my people, representing my people, and working for my people."[42]

According to Eva Perón, she served the *descamisados* with social assistance and love and the workers responded with affection toward her, resulting in her "compenetración espiritual" with her people.[43] Eva offered a model of total self-sacrifice for the good of others:

> I put the soul of my people alongside my own soul. I offer them all my energy so that my body may serve as a bridge leading to a shared happiness. Pass over me, walk firmly, with your head held high, towards the supreme destiny of the new fatherland.[44]

She also sought to be the "puente de unión" between "el pueblo" and Juan Perón.[45] Eva Perón's dedication to Peronism was so manifest that her untimely death from cancer would actually seem to be the ultimate sacrifice, making her a martyr for "el pueblo." She herself suggested this image on more than one occasion toward the end of her life.[46]

The overall goal of the different themes of Peronist spiritual discourse—protection of the nation's Catholic heritage, fulfillment of Church social doctrine by putting social justice into practice, affirmation of the authentically Christian nature of Peronist actions, and promotion of the moral qualities necessary for the individual and the nation—was the utopia to which the wider Peronist project aspired. The idea was:

> to suppress and close forever this fateful cycle of useless struggles between capital and labor; the armed electoral disputes between the political parties;

> among businessmen, by legal or illegal means, publicized or not; to create a different cycle of harmony in which individuals did not fight among themselves like dogs who are thrown a bone; a cycle in which there is harmoniously divided the goods that God wished to sow on this earth in order that we are all happier, provided we are capable of renouncing greed, ambition, and envy.[47]

The plan for accomplishing this was through Peronist doctrine, which was increasingly consolidated and disseminated from about 1948, including the 1949 constitutional reform that wrote the Peronist perspective into the law of the land.[48] The doctrine even merited its own unique name:

> Justicialismo: a doctrine whose objective is man's happiness in human society through the harmony of the material and spiritual, individual and collective forces, enshrined by Christianity.[49]

Caimari has noted that consolidation of a Peronist doctrine diminished the need for Catholicism as a legitimating ideological context. Correspondingly, Perón referred less and less to Catholic social doctrine as the basis for constructing "la justicia social."[50] Promotion of Church interests—and keeping Catholic figures happy—also became a less pressing concern. However, these elements were still retained in the Peronist project, while affirmation of the authentically Christian source of Peronism and assertion of the need to stimulate good spiritual qualities were strengthened. The end result was what may be termed a Peronist cult of civic spirituality.

The Peronist spiritual vision clearly had the capacity to resonate with Catholics. Thus the bishops themselves noted in a 1951 pastoral celebrating the twentieth anniversary of Argentine Catholic Action that:

> There have appeared ... and appear every day, many ways of apostolizing that, in the first moments, have as their chief objective the attracting of men to the faith and fulfilling of religious duties ... nonetheless there have not been lacking through the constant diffusion of the Church's social doctrine, concerns for social welfare, for technical perfection and inspiring movements for sound public laws, very much in tune with the new orientations in the country's institutional life.[51]

For some Catholic Peronists it would be possible to combine both allegiances. Yet for others, doubts would arise about the means employed by Peronism to carry out its doctrine, followed eventually by distrust of the "justicialista" vision of society itself.

In Córdoba, the extreme difficulty of maintaining the unruly Peronist political alliance led national authorities to intervene in the provincial government during the period June 1947–March 1949. Yet the unstable political situation did not obstruct the interaction of Peronist and Catholic views. For example, a priest[52] who was later militantly anti-Peronist recalls that his pastor instructed him to teach a class to parishioners on the relationship between Catholic social doctrine and Peronist doctrine, which he did willingly. He also recalls that Peronist hostility toward charity institutions run by Catholic elite women in Córdoba and elsewhere did not bother the young priests at all, because society had changed and the state needed to become active in this area. The third national *Semana Social* of the *Juventud Obrera Católica* held in Córdoba in 1947 included study and constructive criticism of the government's Five-Year Plan.[53] Father Rafael Moreno had been very active in Catholic labor organizing in Córdoba and was outraged when religion-based unions were banned in 1945. Yet he argued in 1948 that Perón's visit to the city had "un hondo sentido democrático."[54] And Catholic Action praised the Cordoban chief of police for his responsiveness to its concerns about enforcing high standards of morality for behavior in public places.[55]

But clearly there were also some important differences between other Catholic perspectives and Peronism. In 1947, Catholic Action petitioned the federal *interventor* to disallow the establishment of casinos in the province. This concern later reemerged in *Los Principios'* critical editorials in reference to first candidate and then governor-elect, air force general Ignacio San Martín, a close supporter of Perón and promoter of Córdoba's industrialization. In 1949, Catholic Action and many other Catholic groups reacted strongly against a projected bill in the national Senate that would legalize houses of prostitution, closed since 1936. In addition, two local Catholic-inspired entities became completely constrained in their ability to function. The *Sindicato de Obreros y Empleados Públicos* competed with a Peronist union and proved unable to gain official recognition from the government, while the directors of the *Sociedad de Beneficencia* lost their decision-making ability to the *interventor* of the institution.[56]

The editorial tone of *Los Principios* became less strident in 1947–48 as the newspaper adopted a position of constructive criticism toward the new national and provincial authorities. Important differences of opinion remained over issues such as casinos and the *Sociedad de Beneficencia,* the rate of inflation and number of strikes, and facets of Peronist economic nationalism. The editors likewise complained about government interference in the nation's Supreme Court and restrictions on press freedom.

Yet they moved from their late 1946 condemnation of government interference in the provincial *Escuela Normal Superior* to praise of its reorganized plan of studies in early 1948, particularly the inclusion of classes on religion in the program.[57] The editors were also generally happy about the government's housing and educational initiatives on the national and provincial levels.

The editors of *Los Principios* found some more major issues to criticize in 1949: reform of the provincial constitution to accommodate it to the new national constitution; the ultimately successful initiative sponsored by Peronists in the provincial legislature to dissolve the province's independent professional associations of lawyers, *escribanos* (notaries public), and *procuradores* (solicitors); and the size of the provincial budget, said to be expanding a useless and unnecessary bureaucracy. This last concern may have sparked the municipality's closure of the newspaper December 29–30 for supposed health violations. Opposition politicians suspected the governor's office of involvement in the affair.[58] At the same time, *Los Principios* was under investigation by a national bicameral committee on anti-Argentine activities. The committee's members, who were Peronist, had arrived at the newspaper's offices on December 22 to take control of the books of accounting for auditing. The congressional committee was also investigating other independent publications in Argentina, and it would temporarily shut down Catholic dailies in the cities of Santa Fe and Buenos Aires in January 1950. *Los Principios* itself would be unable to go to print for three days in late January because the committee restricted its access to newsprint until finally approving the newspaper's books of accounting.[59]

The actions of the bicameral committee on anti-Argentine activities may have provided a propitious climate for Governor San Martín to vent his displeasure at the editors of *Los Principios* by closing the daily down in late December. The harassment of the newspaper by both provincial and national authorities also demonstrated that underlying tensions between Catholic and Peronist views actually went much deeper than just disagreements over policy, extending to the very right to freedom of expression.

1950–53: An Emerging Conflict in Values

The conflict in Córdoba between Church and state, Catholicism and Peronism, would actually emerge over the course of several years. Catholic and Peronist values intersected in the March 1950 speech of provincial Minister of Education and Culture Alberto Leiva Castro, which listed various directives on the values and conduct befitting educational inspectors in carrying out their duties. His last point was as follows:

> 7. – No factor of self-interest, neither personal nor political, should so much as appear in the inspection duties, no one is above God, the Fatherland, and the school; for that reason, all efforts, all undertakings, all sacrifices at the service of sacred interests must be found in the person of the school inspector.[60]

The Federation of Catholic Teachers and Professors of Córdoba expressed its satisfaction to Leiva Castro for his recent promotions of teaching personnel, terming his work "abiertamente cristiana" and

> so in agreement with the principles upheld by our national and provincial Constitutions and that respond completely to the orientation stamped on the education and culture of our fatherland, by his excellency the president of the Republic.

After making other remarks, the Federación offered Leiva Castro the enthusiastic participation of its members:

> in order cooperate with his excellency in the government's grandest undertaking, which is the Christian education of the youth, beseeching the Lord on a daily basis that the president's steps be guided along the path of social and Christian justice.[61]

Yet a major difference of opinion between Peronist and Catholic perspectives erupted in Buenos Aires in October 1950. A group of Catholic youths interrupted a Spiritist conference there on October 15 when they loudly challenged the conference slogan, "Jesús no es Dios." Police dislodged the Catholic protesters from the scene and attempted to have them disperse in the street, but some 100 had to be arrested. Meanwhile the conference resumed, and during the course of it a work by Perón was read. Besides front-page coverage, *Los Principios* responded with an editorial noting that while the Constitution guaranteed freedom of worship, it also sustained the Catholic Church and required the president to be Catholic. The editors protested against the heretical excesses of the sect and implied that the government should protect Catholicism from such attacks.[62]

Two days later, Perón spoke to the Peronist multitude assembled in the Plaza de Mayo in an event held on October 17 each year to celebrate their loyalty to him. At that time, Perón read to them the 20 fundamental truths of *Justicialismo,* including number 14: "Justicialismo is a new philosophy of life, one that is simple, practical, profoundly Christian and profoundly humanist."[63] But then Perón left on vacation rather than waiting to greet the eminent Church figure who would arrive October 20 to represent the pope at the National Eucharistic Conference to be held in Rosario in late October. Perón also was scheduled to attend the final events of that conference but sent the vice-president in his stead. Meanwhile, Cardinal Copello of Buenos Aires authorized demonstrations of remonstrance to God to be held in the archdiocese's churches for the blasphemy promulgated by the recent Spiritist assembly, and Catholic activists greeted the papal legate with chants of "Viva el Papa," "Cristo es Rey," "Jesús es Dios," "Viva la justicia social cristiana," and "Viva Cristo Rey."[64] Suddenly reversing himself again, Perón decided to attend the closing ceremonies of the Eucharistic Conference after all. Once there the seemingly contrite president kneeled to offer up a public prayer of gratitude to Christ:

> I give thanks to thee because thou hast had the goodness to inspire us from the very heart of your evangelizing a doctrine of justice and love and because thou hast helped us to steadily realize it on this earth and for this people.... As for me, Lord, I ask no more of you than the necessary enlightenment to continue finding the best paths for my people and the strength which may be necessary to lead them to their most noble destiny....[65]

Perón's prayer was actually more self-congratulatory than humble. Apparently factors such as the solemnity of the event, ecclesiastical satisfaction with the president's unexpected attendance, and the widespread sentiment that there really was not much distance separating Peronist and Catholic values meant that his prayer was quite appreciated by Church figures and did not provoke a reaction from Catholic activists. But Perón most certainly had not relinquished his self-authorized power to proclaim the true spiritual values of Christianity and their identification with Peronism, as was evident in a talk he gave before a Peronist audience shortly prior to the closing ceremonies of the Eucharistic Conference:

> Peronism, which perhaps does not care about appearances but tries to assimilate and carry out the substance of things, is an effective, genuine, and honorable way of practicing Christianity.... We love Christ not just because He is God. We love him because he left something to the world that will be eternal: love between mankind. And we love him for his human dignity and the sacrifice against greed, against selfishness, for the common good of brothers. This is how we Peronists understand Christianity, by trying to practice it, by every day doing a good work for the sake of good work, with no other intention in mind than aiding someone less fortunate than ourselves.[66]

A Cordoban delegation attended the conference, and clearly *Los Principios* had informed Catholic readers about the tumult earlier over the Spiritists in Buenos Aires. One can only imagine what Cordoban Catholics and Catholic Peronists concluded about all the drama of these two weeks. One opinion may have been that while Perón certainly had a spiritual side, it was not necessarily Catholic. Undoubtedly others found that Perón's views and actions in Rosario constituted a true confirmation of the Catholic and Christian roots of his vision.

While Peronism was evolving in doctrine and dissemination, the arena of Catholic activity was also changing. As established in 1931, the Argentine Catholic Action consisted of four sectors tightly under the control of the hierarchy and the clergy, divided only by age and gender (men, women, young men, and young women), and based in the parish. But as time went on it became clear that reform was needed to loosen the constraints of the traditional organization and thereby foster lay involvement and vitality. In Córdoba in 1947, the Catholic Action young men's assembly for the archdiocese was held in tandem with a "Congreso de Jóvenes Católicos" to excite interest among a greater and more diverse number of youths through sports, art, and culture, thereby shedding Catholic Action's spiritual and clerical image. In 1948 Archbishop Nicolás Fasolino of Santa Fe spoke to Catholic Action clerical advisers about the importance of specialized groupings in the organization dedicated to university students, secondary students, professionals, and workers. In Córdoba in May 1951, a federation of Catholic cultural clubs was established under the auspices of Catholic Action but distinct from it to stimulate the formation and coordination of "Ateneos Juveniles Católicos." Its broader goal was to serve the religious, cultural, educational, organizational, social, and recreational needs of the youth of the capital city. Finally in June 1951, the bishops themselves authorized the Association of Professionals, University Students, and Secondary Students as a specialized branch of Catholic Action.

Various groups of Catholic professionals were already in existence in Córdoba, such as the "Corporación de Ingenieros y Arquitectos Católicos" founded in 1950 and others established earlier. The *Liga de Padres de Familia* was instituted in Córdoba in 1953, joining the *Liga de Madres* that was already in existence; a major concern of these organizations—as well as the Church and many Catholics at the time—was to safeguard public behavior and denounce indications of immoral conduct. The *Juventud Obrera Católica* (*JOC*), established in Argentina in 1941, was somewhat of an orphan in relation to the Catholic Action movement because of its emphasis on recruiting young workers. The *JOC* began to manifest signs of rejuvenation beginning around 1953. Writing in *Notas de Pastoral Jocista,* the *JOC*'s senior clerical advisers began displaying less spirituality and greater interest in organizational tasks and the concrete reality of the labor movement. They were joined by a number of young priests who had been educated abroad and were highly motivated to seek reform of the Church in its diverse aspects. One such priest was Córdoba's Enrique Angelelli.[67] It is probable that increased mobilization by activist Catholics from midcentury resulted from both new opportunities available for religious activism as Catholic ranks diversified and a Catholic reaction against the widening "peronization" of society.

Other Catholics were also active who already had definitely anti-Peronist views. The respected journal *Criterio* of Buenos Aires, well-connected to the Church hierarchy, had hearkened to the pope's call for democracy at Christmas 1944. The journal displayed a liberal Catholic perspective in its opposition to the Peronist regime. Floreal Forni notes that this perspective was rarely expressed in direct references to Argentine reality. Instead *Criterio* stressed religious and cultural themes, especially coverage of intellectual topics and the experiences of European Catholics whose social-Christian concerns represented an implicit criticism of the Peronist project. The Christian Democrats of Córdoba dissolved their *Unión Demócrata Cristiana* and formed the *Ateneo Social Cristiano* in 1950, but would not found a political party until 1954. By 1950 some nationalist Catholics had taken distance from Peronism. Father Julio Meinvielle of Buenos Aires, for instance, now found Perón's labor policies potentially revolutionary and implicitly Marxist, the manifestation of a Marxist nationalism that would end in collectivism no matter what the spiritual discourse of Perón and the presence of Catholics in the government.[68] And Catholic sentiment was to be found in both major Cordoban opposition parties, the Democrats and the Radicals, although the *UCR* was still markedly Sabattinista.

Peronist restrictions on politics increased in 1951, with measures such as the continued jailing of political and other figures accused of disrespect to the president, more limits on the political opposition's ability to express itself freely, and closure of *La Prensa,* a conservative Buenos Aires daily, which was eventually awarded to the Peronist-controlled *Confederación General del Trabajo* (*CGT*). In a manifesto made public at its first national meeting, the *JOC* included a call for respect of the fundamental freedoms of conscience, expression and the press, association, and assembly. An attempted military revolt in late September led immediately to the establishment of "the state of internal warfare" in the nation, enabling the president to incarcerate citizens without due legal process. A secret presidential directive of April 1952 shows that fears about other possible conspiracies from a number of diverse sectors of society continued to trouble the Peronist leadership.[69]

At the same time, the evolution and dissemination of the Peronist perspective was proceeding. The masculine and feminine branches of the Peronist Party as well as some trade unions in the province of Córdoba continually offered talks and discussions about Peronist doctrine. But dissemination reached beyond the party and its members. In September 1950, Córdoba's minister of education set up a commission to modify the province's educational offerings by including concepts about the national and provincial constitutions based on *El Justicialismo,* a new book by the nation's minister of technical affairs. He also required school principals to assemble their students at day's end on October 16, the eve of Loyalty Day, and explain the importance of the social gains achieved by the constitutional reforms; summaries of the principals' speeches had to be preapproved by his office. In March 1951 Córdoba became the first province to teach about "Justicialismo" under the rubric of "Civismo y moral" in its education program. The National Congress approved Eva Perón's book, *La Razón de Mi Vida,* as a school text in July 1952. The nation's minister of education announced in November 1952 that at the start of the next school year all the secondary-level establishments under his ministry would use a new plan of studies whose main innovation was coursework to ensure knowledge of "La Nueva Argentina." The term referred to a definitely Peronist conception of national reality and accomplishments.[70]

Juan Perón's project to achieve the unity of thought, spirit, and purpose culminated with the institutionalization of Peronist doctrine as the national doctrine through his Second Five-Year Plan. The president presented the plan to the national legislature in December 1952:

> A government program cannot be a cold listing of intentions to be fulfilled nor of projects to realize, but must also have a doctrine, since a national doctrine is the very collective soul; and from this latter is deeply inspired this Five-Year Plan.... The doctrine of the "Second Five-Year Plan" ... is the Peronist doctrine, principles that shape the soul of the "Second Five-Year Plan" and that has as its supreme goal to achieve the people's happiness and the greatness of the Nation, by means of establishing social justice, economic independence, and political sovereignty. It harmonizes material values with spiritual values, the rights of the individual with the rights of society.[71]

The spiritual goal was achievement of a cult of civic spirituality based on Peronist doctrine, as is evident in Perón's words to educational personnel in April 1953:

> The task of indoctrinating the Nation is a slow and painstaking task, one of abnegation and of permanent sacrifice because to indoctrinate does not mean

> simply to teach but also to inculcate. That is to say, it is not only to impart knowledge; it is also to mold the soul. But in order to mold the soul it is not enough to show, one must persuade, it is necessary to convince. For that reason by creating a national doctrine, we have sought to give to the Argentine people a collective soul, and through persuasion, one that must be slow and perseverant, to instill in the spirit of every one of the people the need to struggle for values that are generally greater than power, wealth or any other superficial thing.... It is necessary to know, penetrate and feel, and on the basis of that to nurture a natural spirit which, incarnated in every individual and within the entire community, goes on to create a mystique surrounding this doctrine, a mystique which indicates the need for that doctrine to be respected and carried forward and, when no longer compatible with the circumstances, to be modified.[72]

In Córdoba the conflict between Catholic and Peronist values became more defined during 1952–53. In December 1951 a newly-formed organization of secondary students called the *Centro Estudiantil Heroico* sent a letter to the capital city's mayor denouncing establishments that presented immoral films and shows. *Los Principios* and various Catholic Action groups supported this action and specifically cited the movie theater Cine Hindú. Peronist legislator Alberto Novillo Saravia brought the matter before the provincial Chamber of Deputies and obtained authorization for a note to the mayor seeking the movie house's closure. The municipality promptly closed it.[73] Alberto Novillo Saravia was the son of prominent Catholic Action activist Lisardo Novillo Saravia and a member of Córdoba's traditionalist Catholic elite; he had spoken up before in the provincial legislature on behalf of Church interests. And Catholic pressure had achieved a certain degree of success in the past in motivating municipal authorities to take action against "improper behavior" in the public sphere. However, by May 1952 Alberto Novillo Saravia had been thrown out of the Peronist Party for lack of party discipline[74] and the municipality was allowing the film, *Bárbara atómica,* to be shown with just a warning statement that it was not apt for minors. This time the mobilization was much more militant. Catholics entered the theater and shouted angrily about the scenes being depicted. Some threw objects at the screen, causing considerable damage. Police arrested 100 people. Yet a week later the movie was again being shown in Córdoba, now with a police guarantee to maintain order. Meanwhile the unrest had spread to Buenos Aires, where the same film was being run, and to San Luis where there were protests against an "indecent" live performance in a theater. The police intervened rapidly in Buenos Aires, where some 100 people were detained. The nation's minister of the interior charged that these Catholic youths were engaging in politics without seeking to avail themselves of the proper channels to lodge their protest. He also noted that in Perón's view, moral behavior was premised on the idea of the exercise of one's own will to evade vice. The minister then issued a communiqué charging that anti-Argentine influence was evident in the foreign religious materials seized during investigation of the film attack. *Criterio* responded with a scathing declaration, pointing out that the moral problem did exist, government authorities had not taken action, and Catholics did not follow foreign, anti-Argentine orientations. The journal also intimated that in this and other government measures one might be able to discern a pattern of persecution against the Church.[75]

Why did the government react so strongly against the Catholic mobilization in May? According to the interior minister, there was a fundamental disagreement between the Catholics' emphasis on enforcing moral purity in the streets and the president's opinion on the subject. Moreover, because of the top Peronist leadership's preoccupation with possible conspiracies against the regime, Catholic activists were now being assessed as potential enemies of the State. Thus the June 1952 top-secret "Plan Político" of the *Peronist Comando General* noted that "lately groups of a pro-clerical tendency have made their appearance, causing disturbances of a minor order, although they should not be ignored and a close and prudent vigilance should be maintained." The list of actions to be taken against the administration's enemies included the following: "No. 9. To foment an active propaganda within the clerical environment in order to demonstrate the profoundly Christian spirit of *Justicalismo* and emphasize the false position of some Catholic groups."[76]

Some eight months later, Peronist authorities from Buenos Aires confronted government ministers from the provinces with the fact that the regime now considered its relations with the Church quite unsatisfactory. According to Erio Bonetto, Córdoba's minister of government at the time, national Minister of Political Affairs Román Subiza convoked all provincial ministers of government to a meeting in early February 1953 in the northern city of Resistencia. Among the items on the agenda were the Church's exemption from taxes, the number of religious sects in each province, and relations between provincial ecclesiastical authorities and the state. It quickly became obvious that Subiza and his two advisers expected an anticlerical report to be given for each province. Bonetto vividly recalls that the Buenos Aires government minister spoke first, giving the expected anticlerical statements. Following him in alphabetical order by province was the government minister from Catamarca, Dr. Moreno, a short, dark, and very frank *criollo* quite different in appearance and manner from the cosmopolitan *porteños*. Moreno started his comments quite simply: "En Catamarca, las elecciones las gana la Virgen del Valle." [In Catamarca, the Virgin decides the outcome of elections.] He noted that it would not be politically viable to challenge the Church in his province, and besides that the Peronist government enjoyed good relations with the Church there. Subiza interrupted him several times. Bonetto of Córdoba came next, trying to give an objective analysis. The meeting then adjourned for a meal, during which time the government ministers agreed among themselves to just give accurate information. Although some episodes of discord had occurred, in the provinces they basically did not see much of a conflict between Church and state.[77] While Catholic agitation was a vexing issue for Peronist leaders at the national level, it was much less so for provincial authorities.

Meanwhile, many Cordoban Catholics still continued to find the mixture of religious sentiment in their Peronist allegiance, or Peronist sentiment in their religious allegiance, quite logical. Father Francisco Compañy had left the parish of Ballasteros to become the chaplain of the provincial government of Córdoba and chief of historical research at the National University's *Instituto de Estudios Americanistas*. Compañy was very impressed that the popular outpouring of emotion at Eva Perón's sickness and eventual death in July 1952 also meant that thousands of working-class men and women had gone to church for the first time. He defended the national Congress's designation of Eva as "Spiritual Chief of the Nation" shortly before her death as simply a means of indicating the moral preeminence that made her a guide to Argentines in their struggles to improve the nation. The priest added, "(her) leadership is second to none other that is legitimately exercised in accordance to divine or human law." For Compañy, "Christianity ... take note ... is the axis around which this new National Doctrine revolves."[78] Speaking in December 1953 on behalf of the university reform bill in the national Congress, Córdoba's Senator Elvira Rodríguez Leonardi de Rosales affirmed Perón's constant defense of the fundamentally Christian orientation of Argentine culture:

> We have seen him kneel before the Lord of men and nations in an exemplary way, and we have heard him speak words of astounding humility, asking God to enlighten him in order to fulfill in this part of the world his Christian mission, such as when he said, in a recent speech, that he was convinced that he had been struggling for that justice and for that charity which Christ preached two thousand years ago.[79]

In August 1954, just three months before Perón publicly accused the Church of intervening in politics, Rodríguez Leonardi supported funding for young people's sports tournaments sponsored by the Eva Perón Foundation by citing the need for integral training in mind, body, and spirit as stated in the general educational objective of the Second Five-Year Plan. She went on to conclude:

> Within the organized community in which we live, a new spirit and a new soul are being forged; we all have a place in the struggle to achieve the perfection of the men of tomorrow, the fundamental goal of the Nation ... I believe ... that with these sports activities day by day there is growing the most beautiful of human virtues: solidarity, and it is the job of a good government to strengthen it.[80]

Even Cardinal Caggiano of Rosario still saw considerable space for fruitful Peronist-Catholic collaboration. In June 1953 he told Catholic Action members in Rosario that as always the lay organization was subordinate to the hierarchy and nonpolitical, but emphasized its role in infusing Christian values in the citizenry and supporting social justice and just salaries for workers. He went on:

> Look at how a great part of our fellow citizens, solicited by the Supreme Government of the Nation, in order to collaborate in the Second Five-Year Plan destined for the common good of all Argentines, are getting ready and endeavoring to lend their support.
>
> Do we not have much to contribute?
>
> The Honorable Mr. President repeats it insistently: "we want good men, we need good men, they must be formed in such a way for the Fatherland to be great."
>
> ...[S]o alright, get on the train .... That is your great, profound, lasting contribution to the Fatherland's greatness ... make of men great sons of God and great brothers of their fellow citizens.[81]

Such an attitude did not stop Cardinal Caggiano from personally pressuring the highest spheres of national government. When Perón wanted to initiate a campaign of political pacification in July 1953, he asked for Caggiano's assistance and that of Cardinal Copello of Buenos Aires. In response, the prelates gave him a list of suggestions including releasing political prisoners, reestablishing the freedom of expression, and ending the politicization of the educational system. Perón was not impressed by their advice.[82]

The founding of the *Unión de Estudiantes Secundarios* (*UES*) in early 1953 would add more fuel to the simmering conflict between Church and state, Catholics and Peronists. The *UES* joined other Peronist-dominated organizations already in existence—those of workers, university students, industrialists, professionals—but here the task was to orient the nation's teenagers. Perón himself became increasingly involved with the new organization, ceding part of the presidential residence at Olivos as a recreational club for its feminine branch. He called on the students to learn to manage their own destinies and be free of prejudice, for virtue would not come through ignorance of vice but rather by knowing and dominating it. The means for gaining such knowledge would be by organizing and running their own affairs within the *UES*, which would also help them develop a sense of solidarity with each other. By early 1954 Perón was talking about establishing *UES* centers in provincial cities like Rosario, Córdoba, and Tucumán. He became more emphatic about developing character in young people through the exercise of free will rather than a hypocritical morality based on superficial images or unthinking conventional morality, neither of which were grounded in a person's own conscience. Perón's larger goal was a new, more realistic type of schooling, and along those lines in the *UES* young men and women would be allowed to be in contact with each other "in order that they begin to know one another and so that they begin to think about that superior morality which is the morality which shapes virtue in men and in nations."[83] Perón's views on morality and the intermingling of the sexes went considerably beyond traditional Catholic thinking, but a more immediate issue would be the fact that education and organizing among the youth had long constituted a key arena of activity for the Church.

1954–55: Full-Blown Conflict in Values

By late 1953 to early 1954, many Catholic activists were restless and anxious to affirm their own identities, values, and projects in what had become a Peronized world. Already on May 1, 1953 Córdoba's Catholic trade unions had openly challenged the Peronist-dominated workers' confederation by celebrating Labor Day separately.[84] An editor of the publication *Notas de Pastoral Jocista* primed the priests acting as *JOC* advisers for their Second National Week of Study to be held in the province of Córdoba in February 1954 with these thoughts: "The task of the *JOC* …is nothing less than to penetrate deeply into the working class in order to bring to there, where the priests cannot get to, the light and life of Redemption."[85]

At that *JOC* advisers' meeting, Cardinal Caggiano heard the priests express dissatisfaction with life under Peronism and openly admitted that there were problems, but he hastened to smooth over their discontent. In his speech to the assembled clergymen, Caggiano instead emphasized the improvements in workers' lives over the past ten years and praised the consequent weakness of anticlerical leftists in Argentina's working class.[86] Córdoba's Christian Democrats did not share Caggiano's upbeat vision of the future. Strongly anti-Peronist, they had become very active by 1953 and proposed the consideration of a wide-ranging seven-point agenda in a meeting in Rosario with delegates from Santa Fe and Buenos Aires.[87] By March 1954 the Córdoba group had developed a declaration of principles and proceeded to found the *Partido Republicano* for a democracy of Christian inspiration. In July 1954 the Cordoban Christian Democrats helped to form a *Junta Promotora Nacional de Partidos Políticos Provinciales de Inspiración Democrática Cristiana* with representatives from various provinces meeting in Rosario. Thus was laid the foundation of the Argentine Christian Democratic Party, which would not be publicly proclaimed until July 1955. Meanwhile, Córdoba had been selected as the site for the *Primera Conferencia Argentina de Abogados Católicos* to be held in October 1954. Among the organizers of the event were Cordoban Catholic Action activist Pedro J. Frías, Jr., son of a former governor, and Lisardo Novillo Saravia, Jr., another son of longtime Catholic Action activist Lisardo Novillo Saravia. Both came from Córdoba's traditionalist Catholic elite. And Argentine Catholic Action was to hold its fifth *Semana Social* in Córdoba in March 1954 on the topic "las clases medias,"[88] which was the forgotten social sector in the Peronist project for Argentine society.

Thus for many activist Catholics—be they nationalist, liberal, social, or more purely integral in Catholic perspective—the Peronist-dominated political culture had become too constrained to carry out their mission of re-Christianizing society. Moreover, "Peronist Christianity" certainly did not reflect their Catholic values.[89] Pedro J. Frías, Jr. later recalled how the Catholic Action laity was ultimately unwilling to comply fully with the Peronist project:

> Despite the existing affinities with the conditions which prevailed among militant Catholics, the laymen troops of the *Acción Católica* did not fail to perceive what there was of mass manipulation and Caesarism in the Peronist phenomenon and to the unconditional support demanded it had three reservations: reservations about a socio-political organization which suppressed the autonomy of the individual and of intermediating institutions; reservations about a social justice which often ran counter to simple justice; reservations about a demagoguery which precipitated class struggle.[90]

From this point of view, the "incondicionalidad requerida" by Peronism meant that their Catholic values—and possibly their very religious identity—were in danger of being subsumed. But what would finally bring the conflict in Peronist and Catholic perspectives into the open?

The spark to ignite the conflict would come from the arena of Catholic activity. The issue was instillment of proper values in the youth. The impetus for action was not just competition with the Peronist cultural project but also the chance to experiment with new, looser or more liberal visions of Catholic activism. And the setting was Córdoba. Events moved as follows. Father Quinto Cargnelutti was a young priest who had begun trying out some new ideas for youth activities in the provincial city of Villa María, where he was assigned to a parish. Cordoban Archbishop Fermín Lafitte took the initiative to bring Cargnelutti to the provincial capital to organize the Church's youth activities there, assigning him to just a few hours teaching in Catholic schools so that the priest could get around to many different places and speak to secondary students about his plans. The idea was to organize a Catholic youth movement that although based on Catholic Action groups and the *JOC*, would also appeal to a wider audience of young people from both Catholic and public schools. The immediate plan was to organize events celebrating the Day of the Student in September, including a parade with floats prepared by the students themselves. The overall goal was twofold: to open the Church up more to young people, and to resist Perón's corrupting influence on the youth in the form of the *UES* and his increasing authoritarianism.[91] Cargnelutti recalls about the latter concern:

> It seemed to us a source of corruption. I don't mean by that sexual corruption. There were, of course, circulating many rumors: people commented about the presents that were given to certain young girls in return for certain favors, supposed orgies in the presidential house in Olivos were alluded to.... But that, for us, was not the most important thing, even if it were true. What worried us the most were the methods by which it was attempted to attract the youth. They were not given ideals or a mission: they were seduced with flattery and bribes. These boys and girls were enticed into a frenzy of opportunism which weakened their youthful spirit. And as Catholics that exasperated us, we couldn't permit it.[92]

In a broader sense, what was at stake was the Catholic role in orienting Argentina's cultural values versus what had turned out to be a usurper offering only secular values.

The other motivating factor for forming what would be called the *Movimiento Católico de Juventudes* was the desire to define a more liberal Catholic social presence in contrast to past Church organizational styles.

The movement would have a more flexible and much less developed structure than did Catholic Action: a committee in each secondary school, parish leaders from the ranks of Catholic Action, and a very basic leadership board. The movement was pitched as a broadly Christian organization, and there were neither qualifications for entry nor the separation of sexes.[93] The new enterprise amounted to a vision and a purpose that had little to do with either Cardinal Caggiano's conceptualization of the Church's youth programs—the clergyman had been a key organizer of Catholic Action when it was first founded in Argentina—or his optimism about how to deal with Peronism. Assisting Cargnelutti were fellow diocesan priests Enrique Angelelli, adviser to the *JOC,* and Eladio Bordagaray, adviser to the Catholic *Ateneo Universitario.* These priests thus covered the three main areas of youth activity.[94] Angelelli and Bordagaray were among the best and brightest of Cordoban young diocesan priests, having been selected to complete their clerical training in Europe.

Archbishop Fermín Lafitte of Córdoba played a crucial role in generating the spark that would ignite the conflict with the Peronist state. By 1954 Lafitte clearly was no longer interested in negotiating with Juan Perón at the elite level to smooth over conflicts in Catholic and Peronist perspectives. Without a doubt, Lafitte was well aware of Catholic anti-Peronist sentiment in the archdiocese, and he probably was apprised of most organized efforts to challenge Peronist authority. While Lafitte certainly was not an activist bishop on social issues,[95] he did have a more open perspective concerning Catholic organizational work than did Caggiano. Lafitte had had an excellent track record on avoiding confrontation with the public sphere since becoming the archbishop of Córdoba in 1927. The archbishop had remained unflappable even during Amadeo Sabattini's gubernatorial campaign and subsequent term of office in the thirties, despite widespread rumors that the Radical *caudillo* would unleash religious persecution in Córdoba.[96] Nevertheless, in 1954 the archbishop gambled on a new Catholic youth movement that was very likely to provoke a reaction from the government. In fact, Lafitte told Cargnelutti that the priest would act as his right hand, doing and saying things the archbishop could not because of his office.[97] Yet Lafitte's sponsorship of the new Catholic youth movement did not signal his break with Córdoba's traditionalist Catholic elite, members of which he had named for years to lead Catholic Action and hold other important positions. As a clearly Church-sponsored anti-Peronist measure and an intrinsically *cordobés* initiative, the *Movimiento Católico de Juventudes* was just the type of activity that individuals from the Catholic elite would have strongly encouraged their young family members to join.

The events of the *Movimiento Católico de Juventudes* in Córdoba September 19–26, 1954 went off perfectly and in direct competition with the Peronist *Fiesta de la Juventud* proclaimed by the provincial government for the same period. On September 19 thousands of spectators viewed a large parade with some 100 floats based on regional, student, or Marian themes, followed by sports, cultural activities, and other events, filling out what was termed the First Catholic Student Week. The Peronist alternative included sports, theater, meals, and *UES* events but was poorly attended in comparison. Furthermore, while the Catholic Student Week was still in progress Cordoban Catholic Action called for massive participation in the procession of the Virgen de la Merced, *Generala* of the Army and Patroness of the Air Force. This procession was also very successful.[98]

The Peronist response to such open defiance of its project would not be long in coming. Already in June 1954 Perón had mentioned his concern over what he termed "infiltrados" promoting dissent especially in the trade unions but also in the wider society. In a speech to trade unionists on September 29 he made an oblique comment to the effect that religion should be practiced outside the labor movement. In the annual October 17 speech, Perón declared that at present there were three kinds of political enemies: politicians, communists, and "los emboscados," the last being those claiming to be apolitical or disguised as Peronists. He went on to criticize professional organizations that went beyond their stated purpose, still without offering specifics. The president also staunchly defended Peronist activities for children and the youth such as those organized by the Eva Perón Foundation and the *UES.* On October 22, 16 bishops and the nuncio met privately with the president, minister of foreign relations, and the minister of education. The minister of education made accusations of clerical interference and disturbances, now giving some specific names. The prelates asked for proof and the meeting ended unsatisfactorily. On Sunday morning, October 24, Perón led a procession of *UES* male and female motorcyclists on an outing touring through the city of Buenos Aires. In the first week of November, Peronist-controlled newspapers started an anticlerical campaign.[99] And in the November 10 speech cited at the start of this chapter, Perón publicly spoke out against Catholic Action, Catholic professional organizations, the Christian Democrats, Lafitte, two other bishops, and 18 priests active in provincial Argentina. Córdoba was singled out for special mention:

> It is in Córdoba where, undoubtedly, the strangest things are happening. This priest Bordagaray, advisor to the Ateneo Universitario de Córdoba, is the one

who says that you have to choose between Christ and Perón. What I am trying to do is precisely to defend Christ's doctrine which, for two thousand years, priests like these have tried to destroy but have not been able to.[100]

The conflict worsened, with the Peronist leadership's animosity directed not just at specific Catholic associations, priests, and activists, but also more broadly at Catholics' very presence in positions of authority and the Church's substantial role in orienting the cultural values of the citizenry. This acrimonious environment would force many Peronist Catholics to definitively choose between allegiance to Perón or loyalty to the Church. Repression in Córdoba was especially severe. Within a few days after Perón's speech of November 10, the provincial Judiciary, the National University, and the provincial *Escuela Normal Superior* had been intervened. Many people working for these institutions would resign or be removed, ridding them of nonliberal members of Córdoba's intellectual elite—earlier purges had already cleared out liberal intellectuals—including many priests who also taught. Peronist priest Francisco Compañy would likewise lose his job at the university's *Instituto de Estudios Americanistas.* Other local authorities would also resign or be fired. On November 16 the national minister of education announced that a "spiritual adviser" would be appointed in primary and secondary schools to "inculcate morality," clearly not with Catholic input. Government authorities soon arrested most of the priests from Córdoba named by Perón in his speech and closed the *Ateneo Universitario Católico.* A letter from the bishops to Perón and another to the faithful on November 23 sought to pacify spirits and achieve a peaceful solution to the conflict, but to no avail. And on December 3, the national minister of education abolished the *Dirección e Inspección General de Enseñanza Religiosa,* which had been controlled by ecclesiastical authorities.[101]

Catholics responded in Buenos Aires on December 8 with a massive procession in honor of the Immaculate Conception of the Virgin Mary. In Córdoba Archbishop Lafitte cancelled the outdoor religious event planned to celebrate the day due to the repressive actions being carried out by the Peronist government there. In short order, the Peronist-dominated national Congress passed a law establishing divorce—the measure that as a Catholic Senator Elvira Rodríguez Leonardi could not bring herself to support—and then another regulating the right of public assembly. The latter measure significantly limited the Church's ability to hold further religious processions outside. In Córdoba the provincial legislature voted to remove the allocations in the new provincial budget for paying instructors of religion in the provincial schools and for subsidizing private schools. A resolution by the nation's minister of education reduced the status of religious and moral instruction as subjects in the schools under his jurisdiction. In late December the national government decreed that government-regulated prostitution could again be established. Meanwhile, Peronist officials in Córdoba and elsewhere were assiduously promoting Perón's conceptualization of sports as an excellent arena for the development of spiritual values in Argentina's youth. Córdoba's government moved to facilitate *UES* recreational activities and the founding of sports clubs for the province's youth population.[102]

The new measures on divorce and prostitution reflected the Peronist government's interest in developing fairer and more modern policies in the areas of family relations and public health. The regime took action because resistance from the Church on such issues was now much less important to it. Moreover, the legalization of divorce and prostitution demonstrated once and for all that Catholic cultural values no longer reigned. Perhaps the Peronist regime was also seeking to shock the hierarchy into exercising stricter control over Catholic activists.

Catholics continued to be targeted in Córdoba and elsewhere in the new year, and in March–May 1955 the Peronist state decisively resumed its campaign to curtail public-sector sponsorship of Church influence in society. The executive branch's decree of March 20 reduced the number of Catholic holy days recognized as holidays across the land. In mid-May the national Congress definitively ended *enseñanza religiosa* in the public schools and also religious institutions' exemption from paying taxes.[103] Roberto Carena, a new Peronist deputy representing Córdoba in the national Congress, resigned his seat on May 9 because the Peronist bloc of deputies had decided to seek constitutional reform to separate Church and state in the supreme law of the land. It was around this time that municipal worker and trade unionist Aníbal Ludueña faced the agonizing decision of whether or not to sign a Peronist-sponsored statement supporting constitutional separation of Church and state. And on May 19–20, the national Congress indeed passed the bill calling for such partial reform of the Constitution.[104]

Conclusions

Catholic Peronists like Elvira Rodríguez Leonardi and Aníbal Ludueña had maintained their allegiance to both Catholicism and Peronism even beyond the public eruption of conflict between Church and state in November 1954. But soon the need to make a public choice between one or the other allegiance shattered the integration of their religious and political identities. They had to face the fact that protection of traditional Church interests had suddenly been stripped from the Peronist spiritual project. At the same time, the Peronist vision still retained its other basic concepts: the need for material improvements to achieve social justice (though no longer presented as derived from Church social doctrine); assertion of the authentically Christian basis of Peronist doctrine and practice; and promotion of the spiritual values needed for the new Argentine and the new Argentina. The end to Peronist support for the Church's substantial role in society was a leap that Perón could make not only to defy Catholic dissenters, but also because the Peronist project already included ideas and initiatives that did not necessarily coincide with Church thinking in a wide variety of areas: education, morality, youth activities, women, the family, religious tolerance, public health, and even class relations. Moreover, there had always been tension between the concept of Peronism as the manifestation of genuine Christianity and the role of the Catholic Church in Argentine society. Yet for people like Elvira Rodríguez Leonardi and Aníbal Ludueña, allegiance to Peronist spiritual values was deeply rooted in their Catholic identity. It was impossible to suddenly separate the two without wounding the individuals involved and perhaps their very loyalty to Peronism.

The city of Córdoba was ideally suited for Catholics to force the issue. There dwelled a Catholic elite and traditional Catholic activists increasingly resentful of Peronist dominance of society, plus groups with a more social or liberal Catholic vision such as the *JOC* and the Christian Democrats. And the archbishop of Córdoba proved willing to oppose the regime relatively early on by sponsoring a new Catholic movement for young people that would directly and publicly challenge Peronist efforts to organize the youth. The Catholic youth movement was also in part conceived to invigorate the Church's model of lay organizing among young people. Later, factors such as a strong military presence on the outskirts of the city, the lack of a large industrial working class, and resistance by Catholics and anti-Peronist political groups would lead to Córdoba's successful final revolt against Perón in September 1955.

But what determined why some Cordoban Catholics became anti-Peronist while other remained Peronist until finally forced to choose one allegiance over the other? Certainly class background was a key factor. The Church had typically dedicated more of its organizational efforts to middle- and upper-class Catholics, and they were more likely to become anti-Peronist. Working-class Catholics had found in Peronism a champion of their material interests, and its spiritual vision emphasized the poor over the rich and powerful. For this group, loyalty to Peronism tended to remain intact longer. In the case of Córdoba the traditionalist Catholic elite was part of what Juan Carlos Agulla has termed the city's "doctoral aristocracy," socially and politically dominant until the community began to diversify after 1918.[105] Members of this elite thus were familiar with pressing demands and exercising power without the aid of Peronism.

Perhaps another critical factor for explaining Cordoban Catholics' differing degrees of loyalty to Peronism was organizational identity, or how important Catholic or Peronist identity was in fulfilling the aspirations of each individual, an identity in which class differences were ultimately though not exclusively influential. For Pedro J. Frías, Jr., a lawyer and a member of a proud, traditional Cordoban family, activism in Catholic Action and other Catholic groups was integral to his identity. In contrast, membership in the *Partido Peronista Feminino* gave Elvira Rodríguez Leonardi, a teacher, opportunities she never would have had otherwise, including the chance to become a national senator. It was only after leaving Peronism that she became really active in the women's sector of Catholic Action in Córdoba. Likewise the main arena for activism by Aníbal Ludueña—a *JOC* member as well as a trade unionist—was the municipal workers' union, where he held several offices. Ludueña lost his union position when Perón was overthrown, and he in fact became much more involved with the *JOC* after he no longer could work in the union. Lídia Torrez de Blangetti had lived with her husband and daughter in construction sites after coming to Córdoba in 1949, until she mounted a successful campaign for a bank mortgage to construct a house. She had become active in the *Partido Peronista Feminino* beginning in 1952. Lídia Torrez stopped going to church when, according to her, priests became involved in politics, although she resumed the practice later. Francisco Angulo was a machinist of humble origins who had come to Córdoba from La Rioja. He became very active in the *JOC* and supported the Church in the struggle against Peronism. But by the 1960s, the *JOC* had withered and Angulo became a Peronist union leader in a motor factory in Córdoba.[106] Quinto Cargnelutti was an energetic young priest whose dedication to Catholicism made him distrustful of Peronism long before the eruption of conflict between Church and state. Francisco Compañy was a somewhat older priest whose belief in Peronism probably helped him get appointed as chaplain of the provincial government and head of historical research at a university institute. Compañy's loyalty to Peronism was very strong, partly because of ideological affinity and partly because of the professional possibilities support for the regime had provided him. For such individuals, the conflict between Church and state in Córdoba during 1954–55 clearly disrupted the process of integration of Catholic religious identity and Peronist political allegiance. Yet the affinity between Peronist and Catholic values would remain, and in the future a significant degree of rapprochement would take place. But that is another story.

Notes

1. Discurso pronunciado por el presidente de la Nación, General Juan Perón en la clausura de la reunión de gobernadores de provincias y territorios nacionales," Nov. 10, 1954, Presidencia de la Nación, Secretaría de Prensa y Difusión, Dirección General de Prensa (hereafter Sec. Prensa y Difusión).
2. Interview, Quinto Cargnelutti, May–June 1989. Cargnelutti recalls coming to the provincial capital in March 1954; the official bulletin of the archdiocese lists him as being named an assistant priest to a parish there on January 18, 1954. *Revista Eclesiástica de Córdoba* 1954, p. 24 (hereafter REC).
3. Congreso Nacional, Diario de Sesiones de la Cámara de Senadores, 1954, vol. I, p. 336.
4. Interview, Elvira E. Rodríguez Leonardi de Rosales, December 5, 1989; Félix Luna, *Perón y su tiempo III: El régimen exhausto* (Buenos Aires: Editoral Sudamericana, 1987), p. 227.
5. Interview and autobiographical text, Román Aníbal Ludueña, May 1989.
6. The most persuasive works include: Noreen Stack, "Avoiding the Greater Evil: The Response of the Argentine Catholic Church to Juan Peron," Diss. Rutgers University, 1976; Floreal Forni, "Catolicismo y Peronismo I, II, III," *Unidos* 14, 17, 18 (1987–88); Lila María Caimari, *Perón y la Iglesia católica: Religión, Estado y Sociedad en la Argentina, 1943–1955* (Buenos Aires: Ariel, 1995).
7. César Tcach, *Sabattinismo y peronismo: Partidos políticos en Córdoba 1943–1955* (Buenos Aires: Editorial Sudamericana, 1991).
8. Sílvia N. Roitenburd, "Identidad nacional y legitimidad en el discurso del nacionalismo católico cordobés (1943–1955)," *E.I.A.L.* 5: 2, 1994.
9. Fortunato Mallimaci, *El catolicismo integral en la Argentina (1930–1946)* (Buenos Aires: Editorial Biblos, 1988), pp. 5–13, 31–34, 43–44.
10. *Renewing the Earth: Catholic Documents on Peace, Justice and Liberation*, ed. David J. O'Brien and Thomas A. Shannon (Garden City, New York: Doubleday, 1977), pp. 35–38.
11. He remained active on what became the Archdiocesan Board of Catholic Action until being named *interventor* of the university. *Boletín Eclesiástico de Córdoba* 1934, p. 204; *Los Principios* 28 October 1943, p. 4; Efrain U. Bischoff, *Historia de Córdoba: Cuatro Siglos* (Buenos Aires: Editorial Plus Ultra, 1979), p. 708.
12. *Boletín Eclesiástico de Córdoba* 1931, pp. 266–67; *Los Principios* 12 December 1943, p. 5 and 14 August 1945, p. 6.
13. Coronel Juan Perón, *El Pueblo Quiere Saber de Qué se Trata* (Buenos Aires, 1944), p. 232.

14. Caimari, 1995, pp. 112–15.
15. The *Junta Renovadora* was a breakaway, nationalist faction of the *UCR* that chose to ally with Perón rather than support the *Unión Democrática* ticket.
16. Luis A. J. González Esteves, "Las elecciones de 1946 en la provincia de Córdoba," in *El voto peronista: Estudios de sociología electoral argentina* (Buenos Aires: Editorial Sudamericana, 1980), pp. 337–59.
17. Tcach, pp. 83–90, 168–71.
18. "That 10th of October 1945 [when Perón was forced out of the government and taken prisoner] the Working Class felt it had lost its benefactor and turned out into the streets demanding his release." Ludueña's parents belonged to the Democratic Party. As a teenager he had run away from home and only returned to Córdoba some ten years later in 1951. Ludueña, interview and text.
19. She was so impressed with how Eva Perón treated each person in line for assistance that she joined the Peronist Party upon her return to Córdoba. Interview, Rodríguez Leonardi.
20. Bonetto became mayor of the provincial town of Carrilobo at the age of 31 and soon went on to hold prestigious posts in the provincial legislature and executive branch. Interview, Oct.–Nov. 1989; *Quién es quién en la Argentina: Biografías contemporáneas* (Buenos Aires: Editorial Guillermo Kraft Limitada, 1955), p. 102.
21. Lídia Torrez was a practicing Catholic to the extent possible in a poor rural setting. As a young woman and until she married she travelled to different *colonias*, teaching private classes even though she herself had not completed primary school. Interview, July 19, 1989.
22. Francisco Compañy, *Eva Perón: La abanderada inmóvil*, 2nd. ed. (Córdoba, 1954), pp. 12, 13, 94.
23. Shortly after the election Lisardo Novillo Saravia was appointed president of the *Flota Aérea Mercante* in Buenos Aires; Maldonado Lara was designated finance minister in the first Peronist government of the province. *Los Principios* 16 March, 1946, p. 1; Tcach, p. 88.
24. *Los Principios* 25 June 1944, pp. 2, 5; 22 January 1945, p. 2; 30 January 1945, p. 2; 15 February 1945, p. 2; 16 April 1945, p. 2; 17 July 1945, p. 2; 26 September 1945, p. 4; 16 December 1945, p. 4; 30 January 1946, p. 4; 12 February 1946, p. 4; 24 February 1946, p. 2; REC 1945, pp. 58–60.
25. "Creemos que una verdadera democracia cristiana como la que propugnamos, el bienestar económico del mayor número y la paz política y social de la República, exijen en primer término que concluya el actual régimen militar con sus medios de coerción—estado de sitio, control de la opinión, etc.—y se conforme al pleno imperio de la Constitución, y en segundo término, la efectiva y progresiva eliminación—mediante la ley y con la garantía de la justicia y el derecho—de la perniciosa mentalidad autoritaria que ha hecho posible el advenimiento de aquél." *Los Principios* 11 July 1945, p. 3.

26. *Los Principios* 20 February 1946, p. 5.
27. Consisting then of Argentina's two cardinals and four archbishops—including Archbishop Lafitte—who were meeting in their regular fall encounter at the time.
28. *La Prensa* 19 April 1947, p. 8; "Discurso del General Perón en el almuerzo de la Confederación de Maestros y Profesores Católicos," Oct. 12, 1946, Sec. Prensa y Difusión.
29. According to information provided to *La Prensa* 19 April 1947, p. 8.
30. Letter provided by Father Hernán Benítez, SJ, the person who presented it to the pope.
31. Caimari, 1995, pp. 116–19.
32. Centro de Documentación Justicialista, *La Doctrina Nacional Justicialista, El Estado de Derecho Justicialista, Los Peronistas, La Iglesia y las Fuerzas Armadas* (Buenos Aires, 1973), pp. 71–73.
33. "Discurso del General Perón sobre el Plan Quinquenal ante dirigentes gremiales, en el teatro Colón," Nov. 25, 1946, Sec. Prensa y Difusión.
34. "Discurso del Gral. Perón en el teatro Colón durante el acto organizado por los docentes secundarios," Aug. 4, 1947, Sec. Prensa y Difusión.
35. "Discurso del General Perón, en el acto de tome de posesión del Secretario de Educación," Feb. 19, 1948, Sec. Prensa y Difusión.
36. "Discurso pronunciado el 26 de febrero de 1947 desde la quinta presidencial de Olivos por L.R.A. Radio del Estado y la Red Argentina de Radiodifusión," *Eva Perón: Discursos completos 1946–1948* (Buenos Aires: Editorial Megafón, 1985), pp. 57–58. (Hereafter E.P. : 1946–48).
37. Catholic opinion by this time was largely in favor of female suffrage while cautioning that interest in politics should not cause women to neglect their true duties in the home and with the family. However, while emphasizing that women's proper place was in the home, Peronist discourse also opened up new perspectives for women as citizens and politically and no longer subordinate to anyone except Perón. Susana Bianchi and Norma Sanchís, *El Partido Peronista Feminino (Primera Parte)* (Buenos Aires: Centro Editor de América Latina, 1988), pp. 48–49, 57–65.
38. "Mensaje leído el 2 de mayo de 1947 en el Congreso celebrado en la ciudad de Rosario por el Sindicato Argentino de Maestros," *EP: 1946–48*, p. 83; "Discurso pronunciado el 13 de octubre de 1948 en la ciudad de Remedios de Escalada, Provincia de Buenos Aires con motivo de un homenaje de los trabajadores ferroviarios," *EP: 1946–48*, p. 289. See also her comments on government medical assistance, "Discurso pronunciado el 5 de diciembre de 1949 en la reunión realizada por el Primer Congreso Americano de Medicina del Trabajo en la Sala de Actos del Ministerio de Trabajo y Previsión," *Eva Perón: Discursos completos 1949–1952* (Buenos Aires: Editorial Megafón, 1986), p. 162.
39. "Discurso pronunciado el 27 de setiembre de 1948 en la Secretaría de Trabajo y Previsión ante el personal de Y.P.F.," *EP: 1946–48*, p. 273.

40. "Discurso pronunciado el 5 de noviembre de 1948 en la Plaza Independencia de la capital de la provincia de San Miguel de Tucumán," *EP: 1946–48*, p. 313. In this same vein, Eva frequently acknowledged God for giving her the chance to help others.
41. "Discurso pronunciado el 5 de diciembre de 1949…," pp. 161, 164.
42. "Mensaje pronunciado el 24 de diciembre de 1947 con motivo de las fiestas navideñas," *EP: 1946–48*, pp. 161–63.
43. "Discurso pronunciado el 10 de julio de 1948 en la Bolsa de Comercio con motivo de una donación realizada para la Obra de Ayuda Social," *EP: 1946–48*, p. 243.
44. "Mensaje transmitido el 24 de diciembre de 1948 desde la residencia Presidencial de Olivos por L.R.A. Radio del Estado y la Red Argentina de Radiodifusión con motivo de las fiestas navideñas," *EP: 1946–48*, p. 363.
45. "Discurso pronunciado el 19 de setiembre de 1949 al inaugurarse un barrio en Vicente López, provincia de Buenos Aires," *EP: 1949–52*, p. 131.
46. See, for example, "Discurso pronunciado el 17 de octubre de 1951 ante el pueblo reunido en Plaza de Mayo con motivo de celebrarse el Día de la Lealtad," *EP: 1949–52*, p. 365.
47. Partido Peronista, Consejo Superior Ejecutivo, *Manual del peronista* (Buenos Aires, 1948), p. 84.
48. The governor of Córdoba, Brigadier José I. San Martín, commented: "No es aventurado afirmar en la nueva Constitución, la presencia de un nuevo espíritu, que anima y vivifica el texto legal y, ante todo, una nueva concepción del hombre y de la vida, de la sociedad y de la Nación." *Los Principios*, 14 August 1949, p. 2. Catholic participation was significant in reforming the Constitution and the provisions concerning education and the family in particular generated the interest and sympathy of many Catholics. Nevertheless, the document very clearly manifested Peronist values above all. Caimari, 1995, pp. 173–77.
49. Quoted in Raúl A. Mendé, *El Justicialismo: Doctrina y realidad peronista* (Buenos Aires, 1950), p. 106. Mendé worked closely with Perón as first minister and then secretary of technical affairs in his cabinet 1949–55.
50. Lila María Caimari, "El lugar del Catolicismo en el primer peronismo," *Sociedad y Religión* 9, (July 1992): pp. 71–72; Caimari, 1995, pp. 178–79.
51. Pastoral Colectiva del Episcopado Argentino con motivo del XX Aniversario de la A.C.A.," *Criterio* 1143, (July 12, 1951): p. 548.
52. Name withheld by request.
53. *Los Principios*, 15 February 1947, p. 5; 22 February 1947, p. 4.
54. Moreno went on to conclude: "After two years of revolutionary efforts, no one can ignore nor deny that the government… is breaking with the outdated models of the past and is pledged to achieving, according to its best understanding, a program of national recovery, social justice, and an affirmation of sovereignty that the people applaud and history will judge." *Los Principios* 24 February 1948, p. 4.

55. *Los Principios,* 7 April 1948, p. 2; 9 April 1948, p. 4; 14 November 1948, p. 4; 26 November 1948, p. 4; 27 November 1948, p. 4.
56. *Los Principios,* 4 September 1947, p. 4; 22 March 1948; 17 April 1948, p. 3; 14 November 1948, p. 4; 18 November 1948, p. 4; 11 March 1949, p. 4; 8 May 1949, p. 4; 13 May 1949, p. 4; 14 May 1949, p. 4; 15 May 1949, p. 2; 19 May 1949, p. 4; 22 May 1949, p. 4; 7 August 1949, p. 4; 8 August 1949, p. 3; 14 December 1949, p. 5; Tcach, p. 175.
57. *Los Principios,* 22 September 1946, p. 4; 3 March 1947, p. 2; 10 March 1948, p. 4.
58. *Los Principios,* 27 March 1949, p. 4; 17 April 1949, p. 4; 5 August 1949, p. 2; 31 December 1949, pp. 1, 6; Tcach, p. 175.
59. *Los Principios,* 23 December 1949, p. 4; 11 January 1950, p. 1; 27 January 1950, p. 1; Caimari 1995, pp. 178, 206.
60. *Los Principios,* 25 March 1950, p. 3.
61. *Los Principios,* 31 March 1950, p. 4.
62. *Los Principios,* 16 October 1950, pp. 1, 4.
63. Centro de Documentación Justicialista, p. 102.
64. *Los Principios,* 19 October 1950, p. 1; 20 October 1950, p. 2; 21 October 1950, p. 1; 27 October 1950, p. 1.
65. *Los Principios,* 29 October 1950, p. 1; 30 October 1950, p. 1; *La Prensa,* 30 October 1950, p. 6.
66. Centro de Documentación Justicialista, pp. 71–73.
67. Emilio F. Mignone, "Si es o no legítimo hablar del fracaso de la Acción Católica," *Revista de Teología* 1: 2(1951): 95–98; *REC* 1947, pp. 133–35, 242–43; *REC* 1951, pp. 131–39; *REC* 1953, pp. 135–36; *Los Principios,* 2 November 1947, p. 5 and 20 October 1950, p. 3; Mons. Nicolás Fasolino, "La A.C. y las Especializaciones," *Criterio* 1039 (26 February 1948): 173–78; "Pastoral Colectiva . . . ," *Criterio* 1951, p. 549; Caimari, 1995, pp. 294–97; Abelardo Jorge Soneira, "La Juventud Obrera Católica en la Argentina: De la secularización a la justicia social," *Justicia Social* 5: 8 (June 1989): 77–81; Abelardo Jorge Soneira, "Notas de Pastoral Jocista," *Revista del CIAS,* July 1989, p. 292.
68. Forni, "Catolicismo y Peronismo," pp. 199–201; Ricardo G. Parera, *Los Demócrata Cristianos Argentinos: Testimonio de una experiencia política,* vol. 1 (Buenos Aires: Editorial Leonardo Buschi, 1986), p. 92; Félix Luna, *Perón y su tiempo: La comunidad organizada,* vol. II (Buenos Aires: Editorial Sudamericana, 1987), pp. 253–55; Caimari, 1995, pp. 338–48.
69. *Los Principios,* 24 July 1951, p. 3; Luna II, pp. 198, 242, 421–24.
70. *Los Principios,* 30 September 1950; 14 March 1951, p. 2; *La Nación,* 17, 18, 22 July 1952; 28 November 1952, p. 1.
71. *La Nación,* 2 December 1952, p. 1.
72. "Discurso del General Perón ante directores e inspectores de enseñanza de la provincia de Buenos Aires," April 24, 1953, Sec. Prensa y Difusión.
73. *Los Principios,* 17, 18, 19, 20, 22, 23, December 1951.

74. Alberto Novillo Saravia had voted with two others on a panel of five reviewers to drop a charge of disrespect against José Manuel Alvarez, Jr., counselor for the poor and minors in the local judiciary, for protesting against a Peronist political event held in the courthouse. *La Nación*, 22 February 1952, p. 3; 5 March 1952, p. 3; 18 April 1952, p. 3.
75. *La Nación* 18 May 1952, p. 3; 21 May 1952, p. 3; 23 May 1952, p. 3; 24 May 1952, p. 4; 26 May 1952, p. 2; 27 May 1952, p. 2; "Declaración," *Criterio* 1165, (12 June 1952): 389–92.
76. Luna, II (documents in facsimile); Luna III, p. 199.
77. Bonetto recalls that the meeting took place in late 1953 or early 1954, but *La Nación* reports it as occurring in February 1953. Bonetto, interview; *La Nación*, 4 February 1953, p. 3; 5 February 1953, p. 3.
78. Compañy, pp. 31, 123–25, 142.
79. Congreso Nacional, Diario de Sesiones de la Cámara de Senadores, 1953, p. 1316.
80. Congreso Nacional, Diario de Sesiones de la Cámara de Senadores, 1954, pp. 562–63.
81. Cardinal Antonio Caggiano, "La Acción Católica, sus derechos y sus deberes ante la Iglesia y la sociedad civil," *Criterio* 1191 (9 July 1953): 501–04; *La Nación* 15 July 1953, p. 1.
82. Luna III, p. 200.
83. *La Nación*, 27 January 1953, p. 2; 4 July 1953, p. 2; 28 September 1953, p. 2; 17 January 1954, p. 1; "Discurso del General Perón ante delegados de la Confederación Estudiantil de Institutos Especializados," 19 January 1954, Sec. Prensa y Difusión.
84. Tcach, p. 227.
85. Osvaldo Ganchegui, "La IIª Semana Nacional de Asesores Jocistas," *Notas de Pastoral Jocista*, vol. 7, Nov.–Dec. 1953, pp. 13–14.
86. He went on: "El peronismo no necesita que seamos peronistas y nosotros no lo debemos ser: pero nosotros necesitamos cumplir con nuestros deberes sacerdotales para con nuestros cristianos y conciudadanos y defender sus derechos y apuntalar los bienes conquistados con los principios católicos y con la vida cristiana que debemos difundir en la masa obrera." Antonio Cardinal Caggiano, "Posibilidades de apostolado en la juventud obrera," *Notas de Pastoral Jocista*, vol. 8, March–April 1954, p. 21.
87. Their agenda was to discuss: the definition of social justice; social and economic reforms; the legal and moral position of the family; implantation and defense of the representative, republican, and federal system of government; public administration and the increasing powers of the state; the need for and status of the political parties; and promotion of a constitutional reform. Parera, I, p. 92.
88. Parera, I, pp. 92–98; Tcach p. 227; *La Nación,* 6 June 1954, p. 2; 19 October 1954, p. 2; REC 1954, pp. 120–21.
89. The term is from Caimari, 1995. See chapters 6 and 7.

90. Pedro J. Frías, Jr., "La Política Argentina entre 1930 y 1960 y los Católicos," *30 Años de Acción Católica: Boletín de la Junta Central de la ACA*, 31: 433–34 (April–May 1961): 26–27.
91. Cargnelutti, interview; Luna III, p. 195.
92. Luna III, p. 196.
93. Cargnelutti, interview.
94. Cargnelutti, interview.
95. Although Lafitte did appoint Angelelli as *JOC* adviser in Córdoba. The latter was definitely interested in exploring and promoting new forms of social apostolate among workers. See Enrique Angelelli, "JOC y Parroquia," *Notas de Pastoral Jocista* vol. 8, July-August 1954, pp. 22–38, including these words: "*La hora de la acción ha llegado*, es necesario sacudir el espíritu de aburguesamiento y de comodidad de nuestros católicos y de nuestras instituciones." Reprinted in *REC* 1954, pp. 69–76.
96. Horacio Sueldo, "Los Buenos Ejemplos," *La Voz del Interior* 3 March 1987, p. 4. Sueldo, long an important Christian Democratic leader, characterized Lafitte as follows: "Aquel hijo de Francia criado en la Argentina exhibió ricas dotes de humanista, pastor religioso y orador sagrado. Y no sólo eso: también pensamiento democrático y en lógica coincidencia, fino tacto en la vinculación con el Estado. Su alcurnia intelectual no tenía a menos abrirse al diálogo sencillo y cordial, en el que se descubría su clara visión del mundo y de la historia. Con él, uno tenía la impresión reconfortante de estar hablando no sólo con un obispo sino con un hombre pleno."
97. Cargnelutti, interview.
98. Tcach, pp. 230–31; *La Nación*, 20 September 1954, p. 3; 27 September 1954, p. 5.
99. "Discurso pronunciado por el General Perón en el acto de clausura del Congreso del Sindicato de Obreros y Empleados del Ministerio de Educación," 23 June 1954, Sec. Prensa y Difusión; Luna III, pp. 202–04; *La Nación*, 18 October 1954, p. 1; 25 October 1954, p. 6.
100. "Discurso pronunciado por el presidente de la Nación…," 10 November 1954. The other priests from Córdoba accused by Perón were: Cargnelutti; Manuel Andreatta, a young priest based in the capital city; José Des López, an older priest who headed a parish in Deán Funes; Rafael Moreno, an older priest who had long been active with the Círculo Católico de Obreros in the capital (see footnote 54); Segundo Olmos, a young priest assigned to the town of Salsacate; and Julio Triviño, who apparently was working at the parish of Villa Cura Brochero. Lafitte, Cargnelutti, Bordagary, and possibly Andreatta were all involved with the *Movimiento Católico de Juventudes*, while the others may simply have made anti-Peronist comments that were reported to authorities.
101. *La Nación*, 13 November 1954, pp. 1, 3; 14 November 1954, p. 3; 17 November 1954, p. 1; 17 November 1954, p. 2; 18 November 1954, p. 2; 19 November 1954, p. 3; 24 November 1954, p. 1; 25 November 1954, p. 1; 27

November 1954, p. 3; 3 December 1954, p. 1; 21 January 1955, p. 4; 29 January 1955, p. 3; 4 February 1955, p. 4; 13 February 1955, p. 3; Tcach, pp. 235, 239.

102. *La Nación* , 9 December 1954, pp. 1, 2, 4; 11 December 1954, p. 4; 12 December 1954, p. 3; 14 December 1954, pp. 1, 2; 15 December 1954, pp. 1, 2; 16 December 1954, p. 3; 17 December 1954, p. 1; 19 December 1954, p. 1; 22 December 1954, p. 1; 27 December 1954, p. 4; 28 December 1954, p. 3; 31 December 1954, p. 1; 8 January 1955, p. 4; "Nuestra contribución a la paz de la patria: Declaración Espicopal denunciando la persecución religiosa en la Argentina," *Criterio* 1240 (July 28, 1955): 524–25.

103. *La Nacion*, 21 March 1955, p. 1; 15 April 1955, p. 2; 30 April 1955, p. 3; 12 May 1955, p. 1; 14 May 1955, p. 1; 21 May 1955, p. 1; "Nuestra contribución a la paz...," *Criterio*, pp. 525–26. In Córdoba a provincial government decree on April 29 suspended religious and moral instruction in provincial secondary schools. Provincial primary schools were not offering the subjects either because the specialized teachers of those subjects had not been able to gain entrance to the schools since the recent start of the academic year. *La Nación*, 30 April 1955, p. 3.

104. *La Nación*, 10 May 1955 p. 2; 13 May 1955, p. 1; 19, 20, 21 May 1955, p. 1; "Nuestra contribución a la paz . . . ," *Criterio*, pp. 526–27.

105. Juan Carlos Agulla, *Eclipse of an Aristocracy: An Investigation of the Ruling Elites of the City of Córdoba*, trans. Betty Crouse (Alabama: University of Alabama Press, 1976), pp. 12–25, 83–89.

106. Interview, Francisco Angulo, March and July 1989.

CHAPTER SEVEN

Labor in Córdoba in the 1960s: Trade Union Consciousness and the "Culture of Resistance"

Mónica B. Gordillo

The purpose of this chapter is to explain the particular characteristics of the new social actors who appeared in Córdoba following the establishment of the automobile firms, Fiat and *IKA* (*Industrias Kaiser Argentina*), between 1954 and 1955, firms that generated an important industrial growth in the city. A major concern is to try and explain the reasons for the special role played by the Cordoban working class during the 1960s and the high degree of militancy displayed by the Cordoban unions at the end of the decade, culminating in their participation in the great social protest, the *Cordobazo* of May 29–30, 1969.[1] In the process, the chapter will hopefully establish some interpretive framework with which to better understand the continuities and therefore the radicalization of the 1970s, when there began to take shape *clasista* positions among Cordoban autoworkers .

Obviously there were multiple factors that intervened to make all this possible. Among those factors, the importance of the special political-economic conjuncture during the military government of Juan Carlos Onganía and its social effects should not be underestimated. However, the hypothesis presented in this chapter is not conjunctural, but rather attempts to explain a longer historical process during which a new trade union tradition and consciousness among the autoworkers were taking shape, with particular characteristics derived, in part, from the period when their unions were first organized. This organization occurred within the context of "the Peronist Resistance"—the *SMATA* (*Sindicato de Mecánicos y Afines del Transporte Automotor*)—representing the *IKA* workers and Fiat unions being established in 1956 and 1960 respectively—which implied a new kind of relationship between the leadership and rank and file given the sacking of the established trade union leaders that occurred in these years. In addition to these factors, there must be taken into account the policies adopted by the automotive multinationals who showed themselves disposed to make certain concessions on wages that, by increasing the buying power of their workers, also provided a model for other collective bargaining agreements that contributed to consolidate the domestic market for automobiles.

In this respect, the establishment of the big capitalist enterprise and the struggle over wages and trade union demands within the firm itself were fundamental in consolidating a trade union consciousness and new union tradition based on more democratic practices and a constant mobilization and participation of the rank and file. The consolidation of this new union tradition was favored by the fact that, although with the election of President Arturo Frondizi in 1958 collective bargaining procedures and the legal recognition of unions began anew (thereby beginning a process of readjustment in the country's power structures and with it the integration of the unions with the state and bureaucratization of the labor movement), in the Cordoban autoworkers' unions this process was never fully consummated. This was also true of the light and power workers (*Luz y Fuerza*), the other leading union in Córdoba in this period. In fact, throughout this period, these unions were acquiring an increasing autonomy with respect to their union centrals, being favored by a system of internal organization and the kinds of relations practiced with management.[2]

Starting with the assumption that a trade union consciousness is not the mere reflex of the worker's condition but the result of a complex web of relations between different social actors and of a particular historical experience, this chapter presents the fundamental connection between two levels of analysis: that which makes reference to the function of the union and trade union struggles as they relate to the realm of production, thereby influencing the formation of a union consciousness and a specific trade union tradition, and that which emphasizes a more broader cultural dimension. The latter refers to the meaning of the symbolic, of the values, motivations, and interpretations of reality that cannot be always classified by what is commonly understood as "ideology." These combined influences go on to comprise what I propose is the development of a "culture of resistance" in the 1960s, with characteristics special to Córdoba. Such a culture took shape as a result of a confluence of manifold influences and discourses, among which the "Peronist Resistance" may have enjoyed a preponderant though not exclusive place, and also drew on the internal debates then taking place within the Argentine left.

What is unique about Córdoba in the decade under study was the strength of the antibureaucratic current within the trade union movement that, in many respects, masked a longstanding anti-*porteño* sentiment. This current would be expressed in essentially two distinct tendencies. The first was a more democratic one expressing a leftwing minority tradition and represented in *Luz y Fuerza*, in other "independent" unions allied with the light and power workers, and in opposition groups found in a number of unions. The other, rooted in Peronism's "orthodox" tendency, was the heir of the militant tradition of the first phase of the Peronist Resistance that was crystallized in the declarations of the Peronist labor movement at La Falda (1957) and Huerta Grande (1962), both drafted at trade union conventions taking place in these two Cordoban localities. It should be emphasized that this group can be regarded as antibureaucratic not because these unions were democratic but because they were *verticalista*, that is, they questioned the project of the metalworkers' union leader, Augusto Vandor, not because of his drifting away from the rank and file but because of his interest in breaking the *verticalista* structure of the Peronist labor movement by seeking to institutionalize Peronism in the political system.[3]

In this chapter I thus am interested in analyzing how certain imaginaries and trade union practices were articulated, despite coming from different political-ideological tendencies, in the workers' consciousness; to study the worker discourse in the factory in order to understand how these discourses contributed to strengthen trade union militancy. These discourses made possible the alliance of different sectors, such as in the *CGT de los Argentinos* led by Raimundo Ongaro, which legitimized the working class as a leading social actor by giving it a transforming mission to fulfill in society as a whole. This double articulation allows us to see that although the world of work and the experience of struggle over wages and other trade union demands were fundamental in the constitution of a trade union consciousness, the radicalization of the local working class at the end of the decade must be understood within the framework of this "culture of resistance." The military coup d'etat of June 28, 1966 acted as a catalyst, a precipitator of the tendencies and the radicalized positions that flowered toward the end of the 1960s but in no way was the conjuncture the original cause of the alternative proposals that were launched from Córdoba, but rather these proposals were taking shape throughout a particular historical process that it is necessary to analyze.

The Leading Cordoban Unions: The Problem of Union Organization

In order to understand to what extent a new union tradition was established in Córdoba among the autoworkers and the importance of company policies in the development of that tradition, it is necessary to consider briefly the subject of union organization. The *SMATA* was organized at the national level in June, 1945, obtaining its legal status (*personería gremial*) in November 1947. At that time, its membership was basically comprised of service station mechanics, workers employed by auto dealerships or those working in the few assembly plants and for parts manufacturers; a large automobile industry still did not exist in the country.[4] In Córdoba, however, the *SMATA* local was not established until 1956, immediately following the arrival of *IKA,* and that same year the first collective bargaining agreement was reached. In terms of the local's internal organization, as with the case of the *Unión Obrera Metalúrgica* (*UOM*), it belonged to a centralized industrial union such as those established by law under the Peronist governments (1946–55).

The *SMATA*-Córdoba's membership was spread out among several different companies: the *IKA* factories, *ILASA*, and the Perdriel tool and die plant were all under *IKA*'s authority. In addition, the Transax (Ford) and Thompson Ramco workers were affiliated to the *SMATA*, as were the Fiat workers in the Grandes Motores Diesel plant from 1966 on. The other Fiat plants had not been assigned to the *SMATA*'s jurisdiction since Fiat's original investment in 1954 had been devoted to the manufacturing of agricultural machinery and thus its workers had been assigned to the *UOM;* after the establishment of the *SMATA* local, the autoworkers' union had requested permission to affiliate the Fiat workers, but the request had been denied by the company. The dispute over affiliation of the Fiat workers would become an ongoing source of friction. When Fiat began automobile manufacturing in 1960, the *SMATA* again solicited the right to represent its personnel, but it was then that the Italian company proposed the creation of plant unions, working through certain non-Peronist union activists opposed to the leadership of the Córdoba *SMATA* whose president at that time was the Peronist, Elpidio Torres. It must be emphasized that despite the geographic concentration of the Fiat plants, the proposal was not to create company unions but rather plant unions that were to operate independently and without contact among the members of other plant unions. Thus were organized four unions: *SITRAC* (*Sindicato de Trabajadores de Fiat Concord*), *SITRAM* (*Sindicato de Trabajadores de Fiat Materfer*), *SITRAGMD* (*Sindicato de Trabajadores de Grandes Motores Diesel*) and *SITRAFIC* (*Sindicato de Trabajadores de Fiat Caseros*).[5]

Despite this turn of events, the desire to join the *SMATA* remained strong among a significant number of Fiat workers. This had been demonstrated as early as 1958, when the personnel of the GMD factory requested affiliation to the *SMATA* and, then again, in April, 1965 when an assembly of the *SITRAGMD* union voted to renounce its *personería gremial* obtained the year before in order to solicit affiliation to the autoworkers' union. The "great strike" of July of that same year nonetheless pitted the company against its plant unions because of the Fiat's refusal to accept the requested wage increase and led to the firing of the principal union leadership and the virtual cessation of union activities within the factories. Only the GMD workers, the most skilled of the Fiat labor force, managed to acquire subsequently membership in the *SMATA,* as it did in September 1966.

The light and power workers' union, *Luz y Fuerza,* represented the personnel employed by the *Empresa Provincial de Energía Eléctrica de Córdoba* (EPEC). Unlike the cases of the *SMATA* or the *UOM, Luz y Fuerza* was by Argentine labor law part of a federalist union structure, a "first level" organization affiliated to a "second level" one, that is to the light and power workers' union central, the *Federación Argentina de Trabajadores de Luz y Fuerza* (*FATLYF*). This status gave it a great deal of autonomy in certain areas, as for example in the handling of the union dues paid by its members and on strike measures that, although they had to be communicated to the *FATLYF,* could be carried out without the latter's approval.[6] This greater degree of autonomy made communication more fluid between the leadership and the rank and file and permitted a higher degree of participation for the latter in union decisions.

We see then that the unions analyzed represented three distinct models of formal organization and therefore in the kind of relations existing between them and the national leadership of the union centrals in Buenos Aires: centralized industrial unions, plant unions completely independent of any union centrals, and decentralized unions with great autonomy as regards their internal functioning. But beyond their formal organization, it is interesting to note the real behavior of these unions that, as we will see, displayed as a common trait the high degree of autonomy in their internal administration and with respect to the crucial problems affecting their memberships, characteristics that must be weighted heavily in trying to explain the militancy of the *SMATA* and *Luz y Fuerza* during the decade under study.[7]

Union Behavior and Trade Union Consciousness

The Autoworkers Unions

In order to understand what are the new influences which will contribute to the formation of a new trade union consciousness and working class tradition as they are forged at the point of production and in tandem with the establishment of the union and rank and file participation, it is first necessary to consider the union antecedents of the membership. The majority of the personnel of the automobile companies was comprised of young men between the ages of 21 and 25 at the time of their employment, a characteristic that, combined with the fact that many were migrants to the city, allows us to assume that in the majority of cases this was their first experience with factory life and that they came from backgrounds in which unions had been of scant significance.[8] The only exceptions were an important group of skilled workers who came from the nearby military factories of the *Industrias Aeronáuticas y Mecánicas del Estado* (*IAME*) where they had been subjected to a workplace environment similar to that experienced in the auto plants and where they would have had some exposure to union activities. However, it must be noted that the special characteristics of this factory did not make possible the consolidation of a true union tradition there.[9]

To approach the complicated subject of union consciousness, there are two areas that will be explored. First, management policies and the perception of the union in dealing with them, in order to explore the issue of the "integration" of the labor force into the company and the hypothesized existence of a labor aristocracy with a corporate mentality supposedly characteristic of workers in technologically-advanced industries. Second, we must consider how the union functioned in terms of its relationship with the rank and file.

With respect to the first issue, it is necessary to make clear some differences between *IKA* and Fiat. The North American company transferred to Argentina practices that it had already employed in its country of origin such as accepting the existence of strong union that thereby forced the company to respect the procedures resulting from a union representation of its labor force. Wage commissions, internal grievance procedures, shop stewards and their committees, and other institutions of union life had to be respected, which would in turn buttress the position of the Cordoban local since the company also supported the decentralization of collective bargaining in order to better adjust to market fluctuations. This policy, which enhanced the autonomy of the *SMATA*-Córdoba with regard to its union central, also served to consolidate a new trade union tradition in the Cordoban union by building an experience of mobilizations and direct negotiations in order to achieve their union demands.[10]

Whereas the attitude of *IKA* was an open one, preferring wage increases while guaranteeing strict control of the labor process, Fiat adopted a more paternalistic approach, replicating also the established model of labor relations employed in its country of origin, which tended to maintain low wages in the union contracts though with an important wage incentive, the "premio de la productividad." The latter, a piecework scheme, also in some ways gave the sensation of increasing worker control over the production process. Fiat sought deliberately the integration of the workers to the firm and carried out a gamut of social activities in order to instill in the labor force the idea that they were a part of the Fiat "community." In this respect, though many Fiat workers recognized that *IKA* paid better wages, Fiat appeared to be giving greater benefits to its workers. Thus, for example, the company organized the *Centro Médico Fiat*, the *Centro Cultral y Deportivo Fiat*. Similarly, at the beginning of the academic year it distributed the school uniforms as well as school supplies for the children of all its workers and, during the holiday season, distributed Christmas gifts and held a Christmas party in each of its plants. The wives of the company's upper management played a conspicuous role at these events, generally appearing at their husband's side and in charge of distributing the gifts as a symbol of the close relationship with the workers. Fiat's policies served not only to hide the existence of conflicts through a paternalist posture, but also to prevent any attempt to oppose its authority, thus fomenting the Fiat worker's isolation from the local labor movement and thereby forcing its workers to concentrate on specific work-related problems.

With respect to the issue of integration into the firm, a policy deliberately promoted by Fiat, it is notable the extent to which the Fiat union leadership accepted this concept as revealed in their attitudes toward the rest of the Cordoban labor movement and in the discourse directed toward their own rank and file.[11] In terms of the formation of a trade union consciousness, it is also significant how the union leadership reproduced outside the factory, the model of industrial relations established by the company. For example, on one opportunity, the Fiat unions made a donation to the pupils of a local school, distributing school supplies, converting the union into a "an institution of support and protection" for the school during 1968, making arrangements also that the students would receive a glass of milk and a snack during the months of June, July, and August.[12] In these acts, the union was adopting Fiat's charity model as, for example, when the wives of the union leaders appeared in the photographs of such events, assuming the same role as that played by the wives of company management in the distribution of gifts to the labor force.

In the *SMATA,* in contrast, it is apparent that, despite accusations that were made from time to time that the union leadership cut deals with the company, the union always appeared as something distinct and in conflict with management, even if these conflicts were often resolved through negotiations. In the *SMATA*'s discourse, the company was always portrayed as defending its own interests without concerning itself with the fate of its workers; the capitalist enterprise was portrayed as a great octopus that continuously squeezed more and more out of the labor force. This discourse served to magnify in the membership's perceptions the union's conquests and strengthened rank-and-file militancy, a militancy also stoked by the left-wing opposition that always existed in the plants and that, though it directed harsh criticisms against the established union leadership, indirectly affirmed the necessity of permanent struggle and the fundamental role that the union had to play, with a vitriolic discourse against the company and capitalism in general.

Another element to consider with regard to the question of integration to the company—understood as that identity that links the personal fate of the workers to the expansion and well-being of the firm, thereby weakening a sense of class—is the previously mentioned one of how these unions perceived their situation within the labor movement generally. In this respect, and perhaps in this the Peronist political identity of the majority of the rank and file and of the *SMATA* leadership was fundamental, the *SMATA* always favored participation in movements that had as a political objective the necessity of procuring Perón's return from exile.[13] The proscription of Peronism acted as a unifying element that strengthened the political identity of the working class, transcending the different positions held in the occupational structure. In this way, relations with the rest of the labor movement were crucial in the formation of a worker consciousness, which sought a role for the working class within society. Even if it is not possible to generalize about the existence of such a consciousness among all the autoworkers, I would insist on the general development of such a consciousness through daily experience in the workplace, an experience that in many cases became the first step in the conformation of a broader political consciousness. In another work, I have analyzed the *SMATA*'s bread and butter struggles during the 1960s and have argued that there were continuous mobilizations of the rank and file for the quarterly wage adjustments provided for in collective bargaining agreements after 1960, and to protest specific shop floor problems in certain departments, usually working through the shop stewards' commissions that functioned throughout these years.[14]

The use of mobilizations and a permanent resort to the rank and file served to strengthen the militancy of the union members and solidify ties above all between the workers and their shop stewards, who were the most visible representatives of the union and with whom the workers had direct contact. Although for the *SMATA* leadership this made controlling the shop stewards absolutely crucial, the latter were never wholly under *SMATA* president Elpidio Torres's influence but, on the contrary, the various political tendencies tried precisely there to strengthen their presence, and it was the shop stewards from the plants' various shifts who elected grievance commission that on many occasions caused real problems for the Peronist union leadership. It is noteworthy in that respect how in moments of internal conflict, militant positions intending to break the hierarchical structure of the union surfaced.

Another factor to take into account in considering the development of a trade union tradition are the effects of the periodic readjustments experienced by the autoworkers. Indeed, lay-offs or the very perception of the workers about the instability of their jobs in the company and the subsequent willingness to accept other possibilities of employment, would work against the formation of a stable core of workers and unionism.[15] Nevertheless, in the period under study, the years of expansion and consolidation of the Cordoban auto industry, there was a notable stability of employment. Although job stability was one of the union's principal demands, massive lay-offs were not frequent, especially early in the decade, something that probably contributed to providing a good deal of continuity for the union leadership and the shop stewards. In this way, little by little, workers were coming to realize the importance of the union and the need to participate in it. These feelings were undoubtedly strengthened by the fact that throughout the decade, the union machinery functioned normally. Every two years, the authorities of the executive committee were up for election. The elections of shop stewards positions similarly took place on a regular basis and, very importantly, the opposition had the possibility to express itself through pamphletary literature, in union assemblies, and in the elections in which throughout the period under study, opposition slates to Elpidio Torres always were presented.

The Fiat workers, by contrast, from the very beginning demonstrated some differences with regard to this issue of trade union consciousness. Although their unions were similar in terms of the characteristics of the labor force, the role and the perception of the union by the rank and file were quite different. We have already seen that in this regard, the company itself played an important role in this process. Although it is possible to observe in the early years of the Fiat plant unions an identification on the part of the workers with the union leadership, the role of the union was very different since it limited its actions to negotiations purely on bread and butter issues. The opinions expressed by the Fiat workers confirm the existence of a certain apoliticism in these years, even if there was support for specific union decisions. Following the failure of an important strike in 1965, there practically disappeared from the *SITRAC* plant any union activity; the union president resigned and was succeeded by an individual, Jorge Lozano, whose role would become basically that of presiding over a factory commission that responded to the *UOM* (attempting to recover its loss of jurisdiction over the Fiat workers). From that point on, the workers lost interest in participating in the union; neither the shop stewards nor the grievance committees were subject to reelection and union assemblies ceased. In this way there began a real distancing of the leadership from the rank and file and a loss of identification with the union on the part of the workers, a union increasingly associated with a leadership foreign to their interests.[16]

The union leadership's hostility undermined all attempts at a union opposition within the plant, and instilled in the workers a sense of apathy in everything related to union affairs. These feelings prevented the consolidation in the workplace of a true consciousness and trade union tradition that, in turn, kept the rank and file isolated from the internal power struggles of the union leadership. The situation was quite different in the Grandes Motores Diesel plant where Fiat's most skilled workers had a longstanding desire to join the *SMATA* and struggled to that end, finally achieving their goal in 1966. The rejection of the leadership in the other Fiat plants, however, amounted to a rejection of everything related to union activity, a situation that only would be reversed in the early the 1970s when the antibureaucratic sentiment became one of the predominant imaginaries within the Cordoban labor movement. That sentiment would emerge with even greater intensity in the Fiat plants, precisely because the workers there had suffered the greatest abuse by an unrepresentative leadership.

In the case of the *SMATA,* discontent was on the rise from the time of Onganía's coming to power and the implementation of a more aggressive policy of rationalization on the part of the company (subsequently *IKA*-Renault), a policy made possible by the suspension of collective bargaining negotiations and the freezing of wages as decreed by the government. With the labor movement weakened and internally divided, the automobile firms abandoned the practice of negotiating contracts individually with their labor forces, thereby undermining a fundamental source of power for the local unions. These circumstances and other measures adopted by the Cordoban metal-working industrialists constituted assaults on the very legitimacy of collective bargaining as it had been practiced until that point, and presaged a more direct conflict with the workers that would come in 1969.[17]

The Cordoban Light and Power Workers' Union

In contrast to the autoworkers' unions, among the light and power workers there had existed a union tradition previous to the period under study. *Luz y Fuerza* was established in 1944 and since then had followed an internal policy of pluralism and rank-and-file democracy, with the union regarded as the only possible vehicle to advance worker demands. The light and power workers perceived their situation as a privileged one by virtue of a number of gains they had obtained over the course of their history, but also knew that these gains had been thanks to the unity of the union. This union tradition was reinforced by the fact that the relatives of EPEC employees had preference in obtaining employment in the company and through this practice entire families were found working at EPEC.

The fact that the personnel enjoyed a privileged position in terms of labor stability and that wage demands were easily met, allowed for greater freedom of action to devote time to broader political concerns. In this the role played by Agustín Tosco, the union president throughout the period who tirelessly sought to align union demands with those of society as a whole, was fundamental. However, the union supported the student movement, not only because of strong union discipline and respect for Tosco's sympathy for the students' struggles, but also because the majority of the members of the union were middle-class employees, many of whom were students or had family members who were. As a corollary, this same middle-class mentality made them particularly sensitive to a discourse that highlighted attacks against civil liberties, such as the loss of political rights, the lack of free speech, etc., feelings that would become more intense with the authoritarian tendencies of Onganía's government. Moreover, since Tosco was very adept at handling union negotiations on wage and work issues, the identification of the membership with the union leadership and rank-and-file consciousness were strengthened. It is revealing that despite the political differences that existed among the union members, Tosco's slates faced virtually no opposition in all these years. The following testimony of one worker is illustrative of the degree to which union democracy had become ingrained among the light and power workers:

> I remember how once the actions of seven or eight workers who had broken a strike were being judged in an assembly. Tosco had asked that they not be expelled from the union since that would undermine the union's unity, and we went and defended that position, attempting to punish them without expelling them. But the assembly voted to expel them despite the fact that the executive committee had requested the contrary.[18]

In this way, within the context of a "culture of resistance," in *Luz y Fuerza,* mobilized by the union leadership, the politicization of the union was intensified throughout the 1960s, going so far as to eventually favor more radical positions.

The "Culture of Resistance"

In speaking of a "culture of resistance," it is being argued that in this period there existed an ensemble of ideas, values, imaginaries, and discourses that challenged and questioned the existing order and became powerful enough in the popular consciousness to have penetrated social movements, and thus can be considered to have a broad cultural significance. In fact, in contrast to what had happened in other moments in Argentine history, these cultural practices were not limited to a specific group or a minority but, on the contrary, acquired such an influence within society that they eventually became the common property of very different social sectors.

Toward the end of the 1950s, the first signs appear of a rebellious and "denunciatory" culture, one that would grow in intensity in the following decade. During the 1946–55 Peronist governments, the majority of Argentina's intellectual class, one heavily influenced by liberalism, had united against Peronism and its authoritarian tendencies in such things as education, but without clearly defining a unifying idea; by defining itself in opposition to Peronism it cast its identity in essentially negative terms. Peronism's disappearance from the state after 1955 precipitated then an identity crisis due to a "demonization of Peronism as a national curse" (*un hecho maldito*) that failed to recognize the identification felt with it by the popular classes and led critical intellectuals to reject the liberalism espoused by the *Revolución Libertadora,* thus creating a split in the liberal camp.[19]

New imaginaries began to gain a foothold in Argentine society, among them "the morality of commitment", the concept of "dirtying ones hands" rather than engaging in idle talk.[20] The *Revolucion Libertadora's* "betrayal" and the failure of the program of Arturo Frondizi, in which intellectuals had seen the possibility of implementing the ideals of their generation, would create a propitious climate and lend legitimacy to this "morality of commitment," with its first concrete accomplishment a reassessment of Peronism. In addition, the Peronists more than any other group had their own motives for identifying with a discourse that indicted liberalism, so attacked by Perón himself, for whom action was more important than words and who was an advocate in the effectiveness of "realpolitik."

In the midst of the special conjuncture that was the "Peronist Resistance," it became possible for these various influences to come together and to go on to invigorate certain common imaginaries that, beyond the special meaning they held for each individual, encouraged action. Among those actions was one previously noted: the questioning of liberal democracy and the party system.[21] The need for structural change began to be spoken about, and the idea of revolution—with all its affinities with the idea "getting one's hands dirty" and of the "morality of commitment"—began to be accepted by various sectors of both the left and of Peronism, albeit the interpretations of "revolution" varied. These ideas contained a heavy dose of idealist activism ("voluntarismo"), trying to reconcile a humanistic and materialist conception of man that emphasized revolutionary praxis and gave coherency to and made possible this "morality of commitment."[22] It is important to note that these feelings imbued the diverse participatory union practices that later would play a significant role in moments of mobilization. Another idea that became potent in this period is that of a "national liberation" closely tied to the anti-imperialist struggle, that served also to vindicate the identification with the Latin American struggle, an idea that already existed among intellectuals of the 1950s. As a result of all these influences, in the 1960s there also emerged a strong generational identity. In a period of deep changes, to be young was not only to experience a vital stage in life but also gave one a certain role and implied a "commitment" that was supposed to lead to action; the young had a role to fulfill in society, they were to be the instigators of change and the pioneers of a new world.

Anti-Bureaucratic Tendencies within Peronism and Trade Unionism

I will now briefly analyze the principal groups that gave sustenance to, in different ways but with certain common features, the "culture of resistance," since they provide an essential frame of reference for understanding the characteristics that various social protests, and especially the *Cordobazo,* revealed. The proscription of Peronism and Perón's exile caused various internal currents to emerge within Peronism, currents generally with little in common save the recognition and loyalty to the figure of Perón, and with each appropriating the mantle of being the true spokesman of the essence of Peronism. The spectrum was a broad one and ranged from well-defined leftist positions to others markedly rightist in orientation.

In the first years following Perón's fall, precisely during this period known as the "Peronist Resistance," all channels of political participation were closed. As a result, the idea of conspiratorial struggle provided a common course of action, one characterized by a discourse and tactics that exalted violence and in which ideological differences were hard to distinguish. It would not be until the early 1960s, when President Frondizi initiated the integration process of the unions with the state (when the labor movement entered a phase of bureaucratization) and as international events such as the Cuban and Chinese Revolutions made such a deep impact on Argentina, that the different internal currents within Peronism would take shape.

In this sense, the role played by Córdoba and its unions would be of enormous importance in strengthening one of these currents: the antibureaucratic and combative trade union one. It was here that there was germinating a tradition of resistance that would allow left and right to meet in a common course of action. Unlike what happened at the national level, where in 1957 the meeting to "normalize" the *Confederación General del Trabajo* (*CGT*) failed and would not be achieved until 1963, in Córdoba the *CGT* was reestablished in that earlier year. Although it could not properly speaking be considered a regional "branch" of the *CGT*, since there was no central organization to whom it responded, it fulfilled the function of bringing together different trade union organizations, providing an environment in which, in their daily lives and trade union struggles, a new generation of trade union leaders was created. As a corollary, by not depending on a union central, the Cordoban *CGT* would get accustomed to a high degree of autonomy, an autonomy it would jealously protect after the restrictions on the national labor movement were lifted. In the early years, the Cordoban *CGT* was almost exclusively in the hands of Peronists, hardened by the struggle of "resistance," such as the case of Atilio López, president of the *Unión del Transporte Automotor* (*UTA*), who was elected in 1957 secretary general of the *CGT* local. It was this sector that called the national plenary of the *CGT* locals and of the "62 Organizaciones Peronistas" that took place in the Cordoban locality of La Falda in October, 1957. There a platform was adopted that constituted a veritable blueprint for a future government, a platform that became known as the "Programa de La Falda" and that for many years represented the most ambitious expression of Peronist trade unionism.[23]

At the beginning of the 1960s, this sector continued to control the Cordoban *CGT,* especially with the election in 1962 of *SMATA* president, Elpidio Torres, a man who was to be found within the so-called *legalista* wing of local Peronist trade unionism. Despite the *legalistas* sympathy for Vandor's project and that of their rivals, the *ortodoxos,* for *verticalista* positions, both managed to preserve a high degree of independence from the trade union leadership in Buenos Aires. Moreover, although Peronists controlled the majority of the Cordoban unions, there were important and combative unions such as *Luz y Fuerza* and the printworkers, unions found mainly within the ranks of the so-called Independents, which maintained an ideologically pluralist, left-leaning leadership. Thus, from a very early date there existed what would be the characteristic position of the light and power workers of seeking the establishment of a broad-based front in behalf of objectives that transcended strictly union concerns and fortifying the worker-student alliance their union promoted.

Throughout 1962, and above all as the result of the crisis existing in the metal-working industry (in the midst of declining productivity and profits, leading to employer assaults on union shop floor power and the adoption of rationalization schemes), Córdoba came once again to occupy a leading position in the labor struggles of the period. Its militant position was distilled in a document resulting from a national labor conference held in the province. In July of that year, the so-called *Programa de Huerta Grande,* was released as part of a "mobilization plan" by the "62 Organizaciones Peronistas." The political context was then very hostile to Peronism, the Peronists' gubernatorial electoral victories having just been annulled; and following Frondizi's overthrow in March, any possibility of dialogue seemed closed. It was in this atmosphere that the sector of Peronist trade unionism headed by the textile workers' leader, Andrés Framini, and by the sanitation workers', Amado Olmos, after meeting with Perón published the Huerta Grande document, which revealed a profound radicalization on the part of the labor movement.[24] Nevertheless, this position was not shared by all the union leaders, with a faction already appearing on the horizon more disposed to negotiation and to create structures that would foster a greater independence vis-à-vis Perón, thus breaking with the strict "verticalism" accepted by the *ortodoxo* sector of Peronism.

The rejection of the labor bureaucracy and the negotiating tendency from a leftwing Peronist perspective, was already apparent by 1964 in the euphoria surrounding the *CGT*'s "plan de lucha" of that year and its implications for a massive rank-and-file mobilization and diverse factory occupations over a period of a month and a half.[25] It was then that for the first time appeared the *Movimiento Revolucionario Peronista* (*MRP*), whose program, drafted by Gustavo Rearte, already spoke openly of armed struggle as a tactic and of the necessity of creating a "people's army." In reality, this discourse was not completely new, and picked up on the experience of "the Resistance," though in a version much more ideological and influenced by the preachings of the Peronist Resistance's most leftwing exponent, John William Cooke, a man himself very influenced by the experience of the Cuban Revolution. According to this Castrosit-inspired line, with no other feasible course of action, it was necessary to resort to armed struggle to procure the return of Perón, the only person capable of undertaking the hoped-for transformation of the country.[26] Moreover, Perón himself in a March 17, 1961 letter addressed to the "comrades of Córdoba" stated that "only fools could believe that the reactionaries were going to allow a peaceful transfer of power" and that the working class "had to be the backbone of a decided and energetic action."[27]

However, it must be noted that the revolutionary path was not the exclusive property of leftist groups, but was also strongly rooted in the Peronist political culture of those years. Perón had said that if anyone were capable of realizing the necessary changes without "deeply revolutionizing what already existed, capable of making an omelet without breaking the eggs" that he would be the first to choose that person to lead Argentina. But in the beginning, the idea of revolution was inextricably linked to the cause of Perón's return, to such a point that it appeared to be the revolution's principal objective, without hazarding much, except in the case of the more radicalized groups, about the specific content that this revolution was supposed to have. John William Cooke had already remarked that Peronism was "because of its social composition and its struggles, revolutionary in essence," and clearly identified with the direction that revolution was heading.[28]

Following the failed attempt to return Perón to the country in late 1964, and in light of the endeavors of Vandor's sector of the labor movement to institutionalize the movement, Perón set about affirming the "verticality" of Peronism, sending for that purpose his wife, Isabel Martínez, who strengthened ties with the most *ortodoxo* sectors of Peronism. This maneuver brought to light the differences within Peronist trade unionism, whose outcome was the division at the beginning of 1966 of the "62 Organizaciones

Peronistas" in two camps: the "62 de Pie junto a Perón," who recognized José Alonso, head of the *ortodoxos,* as their leader; and the "62 legalistas" who followed the leadership of the *UOM*'s Augusto Vandor. In reality, there were few important ideological differences between the two groups, but rather a dispute over power. In Córdoba, although the *CGT* pronounced itself in favor of Vandor's faction, the *ortodoxo* camp was quite strong, a fact that would assume importance subsequently when, due to the authoritarian program of Onganía, the intransigent line would become stronger in the city and would permit support for the *CGT de los Argentinos,* a combative anti-Vandorista trade union alliance established in 1968 and led by Raimundo Ongaro of the printworkers' union.

Even though there appeared in Córdoba in 1966 to commemorate October 17 a great number of flyers signed by the "Junta Coordinador de Agrupaciones Revolucionarias Peronistas"—which could lead one to think that there was already some degree of organization—it would only be from 1967 on, after the Onganía government clearly defined its policies, that the contents and strategy of a leftist Peronist revolutionary project were defined with greater precision. In that respect, the internal document released in July 1967 by the *Acción Revoucionaria Peronista,* from which the *Fuerzas Armadas Peronistas* (FAP) would subsequently break away and form, is very instructive. This document is a good example of how the idea of armed struggle, though it did not emerge with Onganía and had been germinating for some time, gained acceptance as a result of the military dictator's policies and made armed struggle a reality.[29] These sentiments nicely complemented the public position assumed by Perón at the death of Ernesto "Che" Guevara, who he eulogized as a true hero, exalting his figure and actions as models to follow by the youth and by Peronists especially, even referring to Che, in very ambiguous terms, as "one of ours."[30]

In 1968, following the initial support offered to the *CGT de los Argentinos,* Perón defined what he termed a "phantom war," guerrilla warfare, as the only possible one in his opinion to confront a militarily powerful enemy. He spoke here of the necessity of inspiring the masses, of "awakening the people's mysticism in order that each group throw themselves into the struggle." His meaning was clear, considering that he was speaking to an audience who already sympathized with such ideas. But Perón also demonstrated an eminently pragmatic sensibility by noting that it was necessary to support all those who were disposed to join the struggle, regardless of their ideological tendencies. He then made clear what he meant by a "phantom war," providing guidelines for what would subsequently become the tactics adopted by the majority of the guerrilla organizations, especially by the *Montoneros.* According to his instructions, it was necessary to give the appearance of leading a normal life and not to expose oneself publicly, to maintain the guerrilla leadership incognito and especially to do it "on a separate string from Peronism's political apparatus."[31] These ideas had their first concrete expression on September 1968, when the *FAP* appeared in the guerrilla actions at Taco Ralo in the province of Tucumán.[32]

The *CGT de los Argentinos* received its greatest support in Córdoba from the city's more leftwing unions such as the light and power workers led by Tosco, a leading representative of the democratic antibureaucratic tendency within the local trade union movement previously discussed. But there also existed the *ortodoxo* current among the Peronist unions, which found common ground with the leftwing unions in their antibureaucratic discourse despite not sharing all the ideological positions of the *CGTA.* This antibureaucratic, antigovernment stand was what generally allowed different social sectors to find common cause in the *CGTA* rebellion.

Peronism's revolutionary current emerged once again in the national plenary held January 11–12, 1969 in Córdoba. Shortly before, in 1968, the "Peronismo de Base" groups had been established, groups that united in one movement the most revolutionary tendencies within Peronism, and which had emerged from or at least had some relationship to the Catholic revolutionaries affiliated with Juan García Elorrio's "Cristianismo y Revolución" magazine and closely tied to Ongaro's line. Part of this group would later join the *Montoneros.* It is very significant that the national executive council of the *Bloque de las Agrupaciones Gremiales y Organizaciones Políticas Peronistas,* precisely the one that sought the alliance between this political wing and the more combative sector of the trade union movement, should have chosen Córdoba as the site to hold the plenary. In fact, Córdoba had become by then one of the most important focal points of trade union resistance to the regime. In the same way, the resolutions adopted to lead the organized struggle "of the working people against their principal enemies: the military dictatorship, the oligarchy, and Yankee imperialism" would serve as a reference point for the social mobilizations that would occur shortly thereafter.

The Non-Peronist Left

Here I will attempt to sketch out in general terms the spectrum of those leftist groups that did not have Peronist origins. It is a difficult task because of the great proliferation of leftwing groups in these years, many of them of ephemeral existence, which emerged in different settings in the 1960s. The problem becomes even more complicated because many of these leftist groups, though they did not have Peronist origins, they would—especially during Onganía's government—attempt to align with Peronism and then in the 70s would assume a Peronist identity upon seeing in Perón's movement the realistic path for a national revolution.

Important international events occurred that would shake up the positions of the Argentine left in the decade of the 1960s. One was certainly the 1959 Cuban Revolution. This revolution would have profound repercussions in all of Latin America, and its example would be regarded as one of the possible paths to obtain "national liberation." Nevertheless, although there occurred some divisions at the party level such as that which took place within the Socialist Party, which split between the *Partido Socialista Democrático* and the *Partido Socialista Argentino* (the latter after 1960 would be called the *Partido Socialista de Vanguardia*), there was not a radical programmatic transformation of the traditional parties of the Argentine left. This, in turn, opened the door for the proliferation of numerous leftist organizations comprised fundamentally of the youth who attempted to define positions in tune with the international changes that were taking place.

In effect, the parties of the left throughout the world experienced a profound ideological debate as a result of the Sino-Soviet rift and the diverse orientations of the "movements of national liberation." Generally speaking, the various leftist organizations in Argentina identified with two broad tendencies: that led by the *Partido Comunista* (*PC*), which continued unwavering in its loyalty to the Soviet Union and as the decade passed was steadily losing popular support; and the second of a Maoist orientation that began as a result of breakaway factions of the *PC* and the *Partido Socialista de Vanguardia.* These dissidents would go on to form the *Partido Comunista Revolucionario* (*PCR*), whose forerunner, the formation of the *PC CNRR* (*Comité Nacional de Recuperación Revolucionario*) in 1967 had already managed to win support among an important group of intellectuals and students in Córdoba tied, to mention just one example, to the group that published the "Pasado y Presente" magazine. In 1972, *PCR* leader Renée Salamanca would win the union presidency of the *SMATA.* Despite Salamanca's brief leadership of the *SMATA,* the communists of both tendencies would have little standing within the trade union movement, save the secondary influence of a third communist tendency, Trotskyism, represented at the time by the *Partido Obrero Trotskista,* which called for rank-and-file direct action, the establishment of factory commissions to establish a genuine worker control over the production process, and permanent revolutionary indoctrination. These Trotskyist positions would subsequently be picked up by the new left *Partido Revolucionario de los Trabajadores* (*PRT*), which in 1970 would exert considerable influence of the new *clasista* leadership of the Fiat unions, *SITRAC* and *SITRAM*.

As previously mentioned, Onganía's government precipitated the emergence of currents that had previously only been latent. Among those was the legitimizing of the "vía armada," of armed struggle. After 1967, there began to emerge a welter of armed guerrilla organizations, each espousing an idea that would subsequently become almost common wisdom within the "new left," that political organizations needed to count on an armed wing to support political activities and social mobilizations. Thus, in 1967, a dissident group of the *PC* merged with sectors coming from the *Partido Socialista de Vanguardia* to establish the *Ejército de Liberación Nacional* (*ELN*), whose objective was to join forces with those of Che Guevara in Bolivia. Following Guevara's death, this plan was abandoned and the group remained in Argentina, giving rise to the *Fuerzas Armadas Revolucionarias* (*FAR*), which after 1970 would join forces with the Peronist left and ally with the Montoneros.[33] In 1968 there was also founded the *Fuerzas Armadas de Liberación* (*FAL*) by *PCR* militants, an organization that would realize the first urban guerrilla action in the country when it attacked an army detachment in Campo de Mayo.[34]

As can be seen, the "vía revolucionaria" had gained prestige among broad sectors of the Argentine left throughout the decade. However, an opportunity was needed for these ideas to find a reception among the masses. These would be found in the various mobilizations that occurred in May, 1969, culminating in the *Cordobazo.*

Radicalization of the Church and the Student Movement

The Catholic Church and student organizations could not escape the process of politicization and the leftward drift of the political culture that the country experienced in the 1960s, and which intensified during Onganía's authoritarian government, especially in Córdoba . In fact, Córdoba occupied a preeminent position in this process, both because of the size and influence of the city's student population and because of the importance that progressive and radicalized sectors within the Church had in Córdoba, such as the "Movimiento de los Curas para el Tercer Mundo." This particular movement gained a national following when some 270 priests endorsed the "Mensaje de los 18 obispos para el Tercer Mundo" at the third world priests' first national gathering, held in Córdoba on May 1–2, 1968.[35] An important turning point within this movement occurred when the newspaper *Córdoba* published separate interviews with three priests, interviews that caused an uproar within the Church and Cordoban society. The three priests were José Gaido, Nelson de La Ferrer, and Erio Vaudagna, each of whom hurled severe—by the standards of Córdoba's conservative Catholic tradition—criticisms against the government. This group, adherents of what would be called "liberation theology," supported from the beginning Ongaro's *CGT*A and contributed to strengthen the idea of a struggle to build a better world, of offering resistance to oppression, of the solidarity in collective actions.

Among the student groups there was also at work a process of radicalization in which the idea of revolution began to grow, dislodging the reformist groups that had heretofore controlled the university. For those who had opted for the revolutionary path, it was the workers, freed from their bureaucratized leadership, who, as the purest and most noble part of society, were to realize this national revolution and transform the country. In this regard, the various student groups in Córdoba were characterized by their attempts to strengthen ties with the workers' movement, in part because they found a sympathetic ear among some union leaders, such as Agustín Tosco. Conscious of the importance of the city's student population, Tosco allowed the light and power workers' union hall on various occasions to be used for student meetings, university prep courses, and similar student-related concerns. Tosco himself, as with Ongaro, on numerous occasions was invited to participate in student gatherings. Moreover, there were some concrete episodes that brought together Córdoba's student and worker populations, such as the marches of the *SMATA* in May 1966 to protest *IKA*'s threat to reduce the workday, when various student organizations went into the street to join the protesters. Similarly, in September of that same year, as a result of the government's intervention of the university, an intense student opposition was unleashed. The disturbances intensified with the occupation of the *Barrio Clínicas* neighborhood and the death of an engineering student and part-time *IKA* worker and union activist, Santiago Pampillón, after security forces were sent to crush the protest. This incident caused great shock and, moreover, by virtue of Pampillón's membership in the *SMATA,* meant that unions were also involved. The September 8 plenary of the Cordoban unions voted in favor of a one-hour strike by each shift to protest repression and violence.[36] After it was learned of Pampillón's death on September 12, the Cordoban *CGT* called for a day of mourning "for all the Cordoban workers."[37] This worker-student alliance would be strengthened with the experience of the *CGT de los Argentinos.*

The Predominant Imaginaries in Worker Consciousness

What I seek to analyze now is the discourse and the practices of the "culture of resistance" in the Córdoba's trade union organizations, to see how the predominant imaginaries were reproduced in worker consciousness, which elements had greater weight at precise historical conjunctures, and to understand the meaning of the actions they occasioned. One way to approach the subject is through an analysis of the flyers and the written material circulating in the factories, produced by both the workers and the various political groups. This pamphletary material allows us to detect the influences exercised in the daily workplace environment that made possible a direct exposure to certain ideas whose impact would grow throughout the decade. Nevertheless, since the first years of the "Peronist Resistance," there had already existed the elements that established the groundwork to make effective this "morality of commitment" that had been gradually influencing certain sectors of the intellectual class since Perón's fall. In this respect, with the establishment of "*CGT* in Resistance" in June 1957, the need to unify the labor movement in the struggle was insisted upon, but it should be emphasized that this objective did not appear to be the result of some intrinsic necessity on the part of the workers, as was the right to protect their material interests, but something inculcated by the paternalistic vision from the years of Peronist government (1946–55).[38]

Nonetheless, with the passing of time, the permanent nature of the *caudillo*'s exile was becoming more of a reality and Perón's return an increasingly remote possibility, while at the same time the political conjuncture seemed to be changing with the opening promoted by Frondizi. This would lead one sector of the labor movement to moderate its intransigent position. Indeed, despite the incendiary tone of Peronist discourse and the revolutionary proclamations that were emitted, the militant labor movement generally did not go beyond, with some exceptions such as the *Peronismo de Base,* the idea of class conciliation, a position that was consistent with what the Peronist governments of the 1940s and 1950s had promoted. This was clearly shown in the moments of greatest euphoria and trade union mobilization, such as the *CGT*'s 1964 "Plan de Lucha" or during the months of the "cabildos abiertos" in August-September 1964, when *CGT* delegations throughout the country summoned distinct social sectors to consider another "Plan de Lucha." By analyzing the ideas presented at these moments, it is possible to perceive the imaginaries with the deepest roots among the working class. All agreed on the necessity of a "change of structures," though practically speaking this did not imply a true change in the system but rather was an attack principally against a political regime that excluded Peronism.

At no point did explicit positions appear among the various trade union organizations that could be considered anticapitalist. What was emphasized rather was the necessity of worker participation in the decision-making process at the institutional level, of the need for strong and powerful unions to carry out a broad array of social functions in housing, health, education, and other programs. Not even the proposed agrarian reform that was included as one of the planks of the "Plan de Lucha" directly attacked the concept of private property since, although it accused the country's great landowners of being oppressors of the rural classes, it immediately associated the former with the interests of foreign powers, and it was this nationalist denunciation that predominated over the landowners' class origins. Indeed, anti-imperialism was Peronism's strongest component, as much among the unions as with the movement's political wing. There was a considerable degree of wishful thinking in Peronist anti-imperialism, because in speaking of national liberation or of the necessity of a "national revolution," it was taken for granted that the national bourgeoisie would be a key player in that revolution and that the working class would perforce be represented in the government that would emerge from such an upheaval, as if the interests of the national bourgeoisie and those of working class always coincided. According to this vision of things, only imperialism upheld exploitation.

One element that allowed a reconciliation between some sectors of the intellectual class and the working class was a reassessment of the concept of the "people" as something pure, incorruptible, as well as the idea of the "youth" as being a status that would reinvigorate society, as being uncontaminated by the vices of liberal society. With this vision, the alliance between the youth and the working class was not only possible but necessary, and would serve as the basis for uniting the worker-student struggles. The "Movimiento de Avanzada Popular Universitario," in the "cabildo abierto" held in Rosario on August 13, 1964 expressed the sentiment as follows:

> The working class together with the youth is the nation's only moral reserve able to carry out the change of structures.... The *Movimento de Avanzada Popular Universitario*... has the honor of having been the first movement to

occupy university buildings in support of the *CGT*'s Plan de Lucha. For us, it is not *CGT*'s plan but the people's plan.[39]

Worker Discourse in the Factory

A simple glance at the political literature that circulated among the *SMATA* membership allows us to see that from the early years that there were always abundant publications of a Peronism in the "resistance," such as *Rebelión Peronista, Trinchera de la Juventud Peronista de Córdoba, Descartes, Lo Mejor que Tenemos es el Pueblo, Punta de Lanza, Nueva Argentina, Retorno,* and others. There circulated, moreover, publications and pamphlets of other political groups, such as *Voz Proletaria* of the *Partido Obrero Trotskista; Palabra Obrera,* spokesman for the *Partido Comunista;* as well as the *PC*'s *Nuestra Palabra.* In 1967, the Communists' publication, *En Marcha,* also appeared.[40] Besides these, all union members periodically received editions of the city's principal newspaper, *La Voz del Interior,* and the union newspaper, *La Voz del SMATA* (along with an occasional supplement of the *SMATA* union newspaper, the *Mecanito*) which kept them informed about all that was happening in the union movement.

But it was the *SMATA* leadership itself that, while emphasizing the importance of the bread-and-butter conquests of the union, pointed out that these were not in themselves sufficient because, as the mystique of the Peronist Resistance would have it, it was necessary for the worker's movement to undertake a greater mission, one defined by the union leadership in the following terms:

> We workers constitute the ONLY authentic revolutionary force in the country, provided, organized and aware, we fulfill our duty as patriots and citizens. Our obligation with the Republic cannot be limited simply to working and earning a wage. Neither can trade unionists limit themselves to Collective Bargaining. The GREAT STRUGGLE, the task we undertake for our children, we will undertake it with great sacrifice, in the Factory, in the streets, everywhere, at all times and with ALL the weapons at our disposal.[41]

The preceding statement does not mean that union responsibilities in negotiations with management were neglected, since from Frondizi's government on, and at least until 1967, the system of collective bargaining functioned normally and the ministry of labor was extremely active, above all during Arturo Illia's presidency (1963–66). Nevertheless, following Peronism's proscription in 1955, one imaginary that became more powerful over the years was that the unions should struggle politically in order to continue the task that, they believed, had been interrupted with the 1955 coup. This idea was so powerful that, despite certain accomplishments achieved by Illia's Radical government in the economy, the unions conspired to overthrow Illia and search for allies who could help to change Peronism's proscribed status.

After two years of Onganía's "Argentine Revolution," with the dictator's program now clearly defined, the language of the various social sectors opposed to the government became more combative, imbuing the discourse of the period with the unions putting greater emphasis on the need for them to play an active role in society. Within the unions themselves there took place a greater politicization, above all of those groups that opposed the established leadership and urged the adoption of different tactics. The antibureaucratic message began to grow more strident, one that resonated in Córdoba's already established union tradition.[42]

Indeed, following the establishment of the *CGT de los Argentinos* in March 1968, the discourse and union practices became much more militant, with the enemies defined as the military dictatorship, imperialism, and the labor bureaucracy. These sentiments also opened the door for Peronism's revolutionary wing by offering it a propitious environment in which to disseminate its ideas, above all given that the frontal assault against the regime was now being led by a workers' labor central (the *CGTA*), which claimed to represent the majority of the working class. Thus, the discourse of revolutionary Peronism, which until now had been confined to an elite within the movement itself, began to penetrate more deeply into the rank and file whose imaginaries were becoming a part of their consciousness, fomenting mobilizations beyond the actual numbers of its outright supporters. Among the personnel of the auto plants there began to circulate publications like *Bombito* (*de fierro*), or revolutionary Peronism's *Con Todo* which began publication precisely in 1968 and which exalted a "regenerative" violence and emphasized the role that the union leaders had to play in this struggle.[43]

Furthermore, the crisis situation created in Córdoba among the unions in early 1969 and the pretensions of Governor Carlos Caballero to create corporatist "advisory councils," forced the unions to take a stand.[44] Reduced work days, plant rationalizations, and the strictures imposed on once powerful, and effective, trade unions led to continuous labor agitation throughout the first months of 1969. The *SMATA* in particular clashed with employers and the state power that supported the company in several violent confrontations. Caballero's efforts to create a corporatist structure through which to get control of the unions and defuse the potential for greater social conflict was an ingenuous and impractical plan for a labor movement in such a state of agitation, with radical currents germinating within the protest over a deterioration in the workers' wages and union protection. The "62 Organizaciones de Córdoba" announced in the trade union plenary held in February of that year:

> We will not bow to capital, we will not bow to the aristocracy and the oligarchy, we will not serve as their useless idiots. We do not want advice, we do not want influence, WE WANT POWER TO THE PEOPLE. . . . After Buenos Aires, Córdoba is the political center of the country; if in Córdoba we Cordobans were to accept the advisory councils, the entire country would dance to the music and the rhythm played by the provincial government. . . . [T]he Working Class prefers the proud independence of its arduous struggle without quarter.[45]

The previous statement also reflects the belief in the role that Córdoba's trade unions were playing in the country and how the alternatives being played out there could have a decisive impact on the future development of the national labor movement and on society as a whole. This feeling of distinctiveness from Buenos Aires, which drew strength from strong anti-*porteño* sentiments, must be taken into account in order to understand the subsequent social mobilizations that took place in 1969 and the alternatives that appeared in the 1970s.

The Clasista Tendency within the SMATA

When the *SMATA* was first organized, it was a leftist leadership that came to power. After the 1958 election of Elpidio Torres, the union would be in Peronist hands but with a leftist opposition throughout the 1960s. These leftwing groups were generally concentrated in the most skilled departments of the plants. Although they were not numerically very important, the role they played within the shop stewards' committee and their insistence on maintaining a presence in the factories made them a force to reckon with. One of the groups that most stood out in both respects was the "Fracción Trotskista de Obreros Mecánicos," organized as a wing of the *Partido Obrero Trotskista.* In addition to publishing its newspaper, *La Voz Proletaria,* throughout these years, its flyers appeared on the factory floors every time there was a problem in the workplace or within the labor movement. In reality, this was the only group within the union that proposed an anticapitalist position from the very beginning, though one very tied to the dualistic logic, capitalism-imperialism. Although they were a minority, during moments of conflict when the rank and file were especially sensitive, leftist activists loomed as a potential danger to the union leadership by incessantly questioning its handling of the union and by ideas that may have at least indirectly influenced the rank and file. Moreover, these leftwing groups periodically urged supporting dissident Peronist slates, as they did in the mid-1960s with the "Agrupación 18 de marzo," which appeared as an alternative to the established Peronist leadership. The Trotskyist faction considered this particular group to represent the "class and revolutionary tendencies of Peronism" within the factory.[46] The position of the Trotskyists was not free from the wishful thinking of the period, in the sense that they thought the "masses" were already prepared for revolution and that all that was needed was to organize them in order to carry out one. This was clearly demonstrated in their interpretation of the student protests of September 1966, when they proposed forming a worker-student front and, through neighborhood committees, holding a regional congress of factory, neighborhood, union, and student representatives.[47] Regardless of the viability of the Trotskyists' agenda, their discourse served to strengthen worker militancy.

In 1967, the flyers of other political organizations began to appear in the plants, flyers that denounced the idea of class conciliation and supported a genuine anticapitalist struggle. With the *CGT*'s call for a general strike in March of that year, the *Partido Revolucionario de los Trabajadores,* for example, insisted on the need to carry out the strike as part of a "plan de lucha" by both the *CGT* national and the *CGT* local and to organize a *clasista* leadership of the *CGT.*[48] Another group within the *SMATA* that emerged with a position opposed to conciliation with management was the "Agrupación 1 de Mayo," which does not seem to have belonged to a specific party and may have arisen as an attempt to resolve workplace problems in a more confrontational way and through practicing a more direct rank-and-file democracy.[49] In late 1968 there also appeared the group calling itself "Vanguardia Obrera Mecánica." Its activities became more visible with the April 1969 strike of the Perdriel (*IKA*-Renault) plant and in the campaign undertaken to oppose the "extraordinary" union dues contribution requested by the *SMATA* executive committee that same year. This group also signaled the need to organize "clandestine resistance committees" in every department that would follow a mobilization campaign against the revocation of the "sábado inglés" law and to impose worker control over productivity quotas.[50] Finally, with the climate of euphoria unleashed by the *Cordobazo,* there appeared in the *SMATA* the "Tendencia de Avanzada Mecánica" with the explicit intention of establishing a "*clasista* current" in the union that would put an end to the "tomb that the corrupt union leadership represented" and would struggle not only to obtain wage increases but also against the problems of rationalization that were being adopted with varying intensity in different departments."[51]

Throughout 1969, the radicalized positions grew stronger, in part because the *SMATA* executive committee's weakness following the imprisonment of the *SMATA* president, Elpidio Torres, after the *Cordobazo*. This message of the need for a *clasista* leadership combined with the political conjuncture of the early 1970s would make possible the triumph of the *PCR*'s Reneé Salamanca in 1972, though these tendencies did not emerge with the *Cordobazo* and had been taking shape through the 1960s.

By Way of Conclusions

This chapter sought to focus attention on some aspects that must be taken into account in order to understand the special characteristics of the new social actors who emerged in Córdoba during the 1960s, aspects that fundamentally explain the high degree of militancy displayed by the Cordoban autoworkers at the end of the 1960s and in the early 1970s. Of special importance was the question of trade union consciousness and its relationship with the workers' condition as it existed in a leading industrial sector. Only by understanding this consciousness can some sense be made of that working class's subsequent radicalization, as well as assessing whether the concept of the classic Marxist concept of a "labor vanguard" explains this phenomenon or whether instead multiple, specific factors intervened and need to be considered. As part of the latter undertaking, it was thus necessary to consider the function of the union in the formation and consolidation of this trade union consciousness as evidenced in the struggle for wages and working conditions. But at the same time, by virtue of the autoworkers' unions being organized within the framework of Peronism's proscription, the union also came to assume a political role, acting as a conduit through which were channeled, not only bread- and butter-demands, but also the political aspirations that lacked an outlet, a situation that contributed to weakening any feelings of integration with the company.

However, the chapter attempted to highlight the differences existing between those autoworkers affiliated to the *SMATA* and those who were members of the Fiat plant unions, differences that make clear the necessity to take into account the complex web of relations that impinged on the conformation of the workers' consciousness. In this respect, not only is it possible to see important differences in respective companies' productive policies, but also in the qualitative nature of the companies' relationship with their labor forces.[52] There was a great distance between the predominant attitudes of workers in the Fiat plants who were characterized by their apathy and lack of trust in their leadership—attitudes that led to a lack of interest in participating in the union—and those of their leadership who preached, especially between 1965 and 1970, integration with the company, rejecting any kind of antimanagement discourse and seemingly having internalized the idea of the Fiat "family." In contrast, by the end of the 1960s, there already existed in the *SMATA* a consciousness and a union tradition that were consolidated as a result of the mobilizations undertaken throughout the decade. Moreover, by that point the economic and political context allowed significant numbers of workers to assume more radicalized positions that strengthened feelings of a worker identity. These very different trade union cultures help to explain why one sector the of the local autoworkers, those affiliated to the *SMATA,* played a leading role in the May 1969 events, while the other, the Fiat workers, were absent in the *Cordobazo.*

In the radicalization that took place at the end of the 1960s, there were obviously multiple influences. Some of them were exclusive to the workplace and the efforts by management to apply stricter programs of rationalization, whereas others derived from political-economic situation in the country in general. Nevertheless, this chapter has attempted to analyze the influence of the slow conformation of a "culture of resistance" and of an opposition within the context of a deepening militancy of different social actors, and the specific repercussions that this situation had on the discourse and tactics of Córdoba's leading unions. Thus, it can be concluded that there existed the slow conformation of a contestatory culture throughout the decade in which there grew stronger—drawing sustenance from different political tendencies—the imaginaries that emphasized the leading role that the working class was to play, along with that of the youth, as the country's moral reserve and leading protagonist in pursuit of the changes the country needed; these imaginaries caused the struggle undertaken by the workers' movement to appear legitimized and supported by society as a whole, strengthening in the process worker militancy.

Translated by James P. Brennan

Notes

1. On the Cordobazo, see James P. Brennan and Mónica B. Gordillo, "Working Class Protest, Popular Revolt, and Urban Insurrection: the 1969 Cordobazo," *Journal of Social History*, 27: 3 (Spring, 1994): 477–498.

2. Mónica B. Gordillo, "Los prolegómenos del cordobazo: los sindicatos líderes de Córdoba dentro de la estructura del poder sindical," *Desarrollo Económico*, 23: 122 (July-September, 1991): 183–187.
3. *Verticalista* refers to the hierarchical structure of the Peronist movement, which recognized Perón as the supreme authority in all decisions.
4. Gordillo, "Los prolegómenos del cordobazo." This subject is explored in greater depth in this article.
5. *Boletín Oficial de la República Argentina,*16 September 1964, pp. 4–7.
6. *Boletín Oficial de la República Argentina*, 8 June 1963, "Estatutos de la FATLYF."
7. See Gordillo, "Los prolegómenos del cordobazo" where the relationship maintained between these unions and their centrals is analyzed in more detail, especially the issue of union autonomy and how this autonomy was strengthened by the experience of direct confrontation with the company.
8. In Mónica B. Gordillo, "Características de los sindicatos líderes de Córdoba en los' 60. El ámbito del trabajo y la dimensión cultural," *Informe Anual al CONICOR*, Córdoba, April, 1991, I present a statistical analysis of the personnel affiliated to the *SMATA* in which, among other variables, I consider the age at the moment of entering the employment of the company and also place of origin.
9. By virtue of being a military factory, *IAME*'s labor force was subject to a strict discipline in which union activities were tightly controlled.
10. In Gordillo, "Los prolegómenos del cordobazo," I analyze in depth the demands of the *SMATA* and *Luz y Fuerza* throughout the period under study.
11. These sentiments were revealed in a very graphic way during the visit of Onganía's minister of the interior to the Fiat Materfer plant in June 1967. On that occasion, Materfer union president, Hugo Cassanova, said: "Our union has always avoided strike measures because we understand the importance of continuing to work. Just that, to keep working. We've lived from the start in a close relationship with the company, despite moments of disagreement, convinced of the importance that its production meant for Argentina's railroads (the Materfer plant manufactured railroad equipment) and for the country." *Los Principios*, 1 June 1967.
12. *La Voz del Interior*, 13 April 1968.
13. See, for example, the participation of the *SMATA* in the *CGT*'s "planes de lucha," and Elpidio Torres's actions as secretary general of the Cordoban *CGT* and member of the "62 Organizaciones Peronistas" as discussed in Mónica B. Gordillo, "Características y proyección nacional del sindicalismo cordobés, 1960–1966." *Informe anual al CONICOR*, April, 1989, and "Características y proyección nacional de los sindicatos líderes de Córdoba, 1960–1969," *Informe anual al CONICOR*, April, 1990.
14. Gordillo, "Los prolegómenos del cordobazo."
15. José Nun, "Despidos en la industria automotriz argentina: un estudio de un caso de superpoblación flotante," *Revista Mexicana de Sociología*, XL: 1 (January–March, 1978): 55–106.
16. Interview with Raúl Batistella, Fiat-Concord worker, July 4, 1990.

17. The Cordoban metalworking firms refused to accept the suspension of the *quitas zonales*, a practice whereby, for purposes of industrial promotion, the workers in certain provinces received a reduced percentage of the wages agreed to in the UOM's national contracts, a suspension established through government decree in 1966. To this was added a government decree in May 1969, which revoked the practice of *sábado inglés* whereby workers worked a 44-hour week but were paid for 48.
18. Interview with Dante Antonelli, *EPEC* employee and union activist, November 28, 1989.
19. Oscar Terán, *En busca de la ideología argentina* (Buenos Aires: Catálogos, 1986), p. 215.
20. Terán, *En busca de la ideología argentina*, p. 201.
21. See Gordillo, "Característiacs y proyección," pp. 72–76.
22. From the intellectuals' perspective, these ideas were highly influnced by the Gramscian idealism that was so strong in Córdoba, best represented by the group of intellectuals who published the magazine, *Pasado y Presente*. See Oscar Terán, *Nuestros años sesenta* (Buenos Aires: Puntosur, 1991), p. 104.
23. See Roberto Baschetti, *Documentos de la Resistencia Peronista, 1955–1970*, (Buenos Aires: Puntosur, 1988), p. 66.
24. See Baschetti, *Documentos de la Resistencia*, in which the complete text of the Huerta Grande program appears.
25. See Gordillo, "Características y proyección nacional de los sindicatos líderes de Córdoba, 1958–1969," Ph.D dissertation, Universidas Nacional de Córdoba (1993).
26. For the MRP, the workers had to shed their bourgeois leadership and convert the Peronist movement into the leader of a revolutionary process. In these years, there was as yet no thought given to a break with Perón (as there would be in the 1970s), and it was felt that the *caudillo* should lead the movement. See Gil German, *La izquierda peronisa (1955–1974)* (Buenos Aires: CEAL, 1989), p. 45.
27. Letter from Perón to "the Comrades of Córdoba," March 17, 1961, in the archive of the *SMATA*-Córdoba (henceforth ASC).
28. Perón said "we've already entered the stage in which not a bourgeois nationalism but a social revolution and national liberation are not separate objectives but rather two facets of the same indivisible process..." These words appeared in an article entitled "Definiciones" published in *Cristianismo y Revolución*, 2–3 (October–November, 1966).
29. In this internal document, the *Acción Revolucionaria Peronista* interpreted the situation in the following terms: "The class nature of the regime remains unchanged but its superstructural composition has suffered modifications.... Our strategy is today, as always, that of armed struggle." in R. Baschetti, *Documentos de la Resistencia Peronista*, pp. 238–239.
30. Letter from Perón to Peronist Movement, Madrid, 24 October 1967, in Baschetti, pp. 273–274.
31. "Mensaje," September, 1968 in *Cuadernos de Marcha*, 71, (June, 1973): 16.

32. "Comunicado ded Destacamento '17 de octubre' de las FAP," November 1968, in Baschetti, pp. 297–299.
33. Oscar Anzorena, *Tiempo de violencia y utopía (1966–1976)*, (Buenos Aires: Contrapunto 1988): pp. 83–84. According to this author, the *PRT* emerged in 1963 as a result of the merging of the *Frente Revoucionario Indoamericano Popular* and the *Palabra Obrera*.
34. Anzorena, pp. 53–54.
35. B. Balvé and B. Balvé, *El 69. Huelga política de masas. Rosariazo-Cordobazo-Rosariazo* (Buenos Aires: Contrapunto, 1989, p. 98.
36. Resolution of the "Plenario de Gremios Confederados de la CGT Córdoba," September 8, 1966 in ASC.
37. Circular of the "Comisión Directiva del *SMATA*," September 15, 1966 in ASC.
38. Thus, as a justification for reestablishing a union central, it was stated, "That it is necessary to make uniform the campaign of a slow restitution for Peronist trade unionism which we have undertaken, not as an end in itself, but as one of the means to realize the fundamental objective of the Peronist movement: the return of Juan Perón." In Baschetti, p. 58.
39. Recording of the "cabildo abierto," Rosario, August 13, 1964, CGT archive, Buenos Aires.
40. All these publications, save the last, are from the years 1961–1965 and can be found in the *SMATA*-Córdoba's archive.
41. ASC, speech pronounced by Elpidio Torres, May 30, 1963 in the commercial workers' union hall during a public demonstration called by "62 Organizaciones," "Gremios independeientes," and the Cordoban CGT. The "24 de febrero" group reproduced parts of the speech and had them distributed to the *SMATA* membership.
42. The belief in a distinct Cordoban trade union tradition, different from that of Buenos Aires, remains widespread among rank-and-file workers. Some representative testimonies are that of Raúl Batistella, a Fiat Concord worker, who stated, "I don't know if Córdoba's leadership was more honest, but it was more controlled by the rank and file," and that of IKA-Renault worker, Humberto Brando, who asserted that in Buenos Aires, "there was the labor bureaucracy, in Córdoba no"; Interviews July 4, 1990 and July 12, 1989 respectively.
43. "Violence is man himself becoming whole, is man himself exercising his right to obtain his title to humanity.... We have learned that our gun will be our title to humanity, our title to citizenship, not laws, not codified rights, not constitutions, not statutes." *Con Todo. Organo del Peronismo Revolucionario* 5 (February, 1969).
44. The crisis I refer to is the conflict over the *quitas zonales* mentioned previously.
45. Published by the "Mesa Coordinadora de las 62 Organizaciones," Córdoba, March 5, 1969.
46. ASC, flyer of the *Fracción Trotskista de Mecánicos*, 17 September 1963.
47. ASC, flyer of the *Comité Regional Córdoba del Partido Obrero*, Septmber 8, 1966.

48. ASC, flyer of the *Partido Revolucionario de los Trabajadores*, March, 1967. The flyer's last words were; "For a revolutionary government of the workers and the popular and workers' parties."
49. ASC, flyer of the *Agrupación 1 de Mayo del SMATA*, October 14, 1968.
50. ASC, flyer of the *Vanguardia Obrera Mecánica,* May 5, 1969.
51. Publication of the *Tendencia de Avanzada Mecánica,* 1 (July, 1969).
52. James P. Brennan, "El clasismo y los obreros. El contexto fabril del 'sindicalismo de liberación' en la industria automotriz cordobesa, 1970–1975," *Desarrollo Económico*, 32: 125 (April–June, 1992): 11–13.

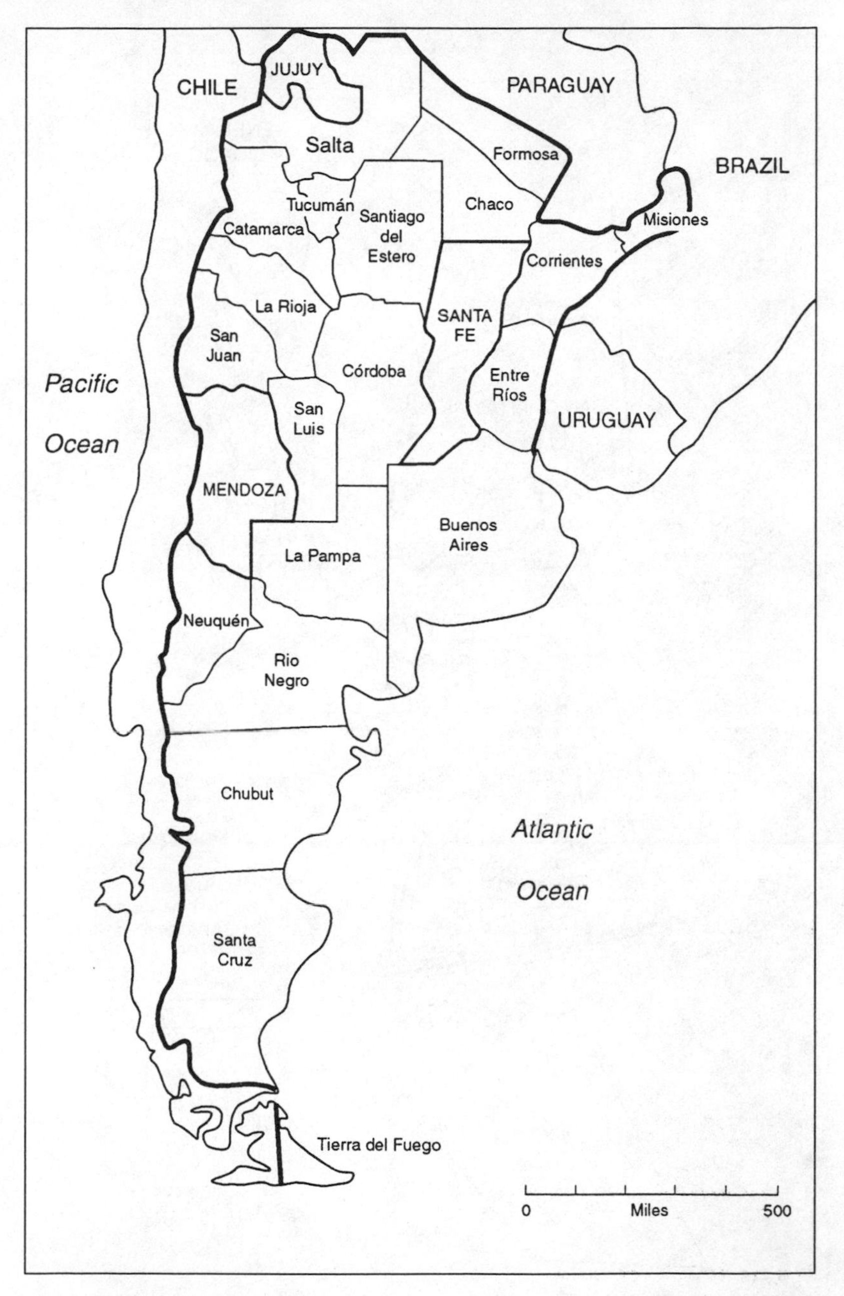

Map I Argentina and its provinces

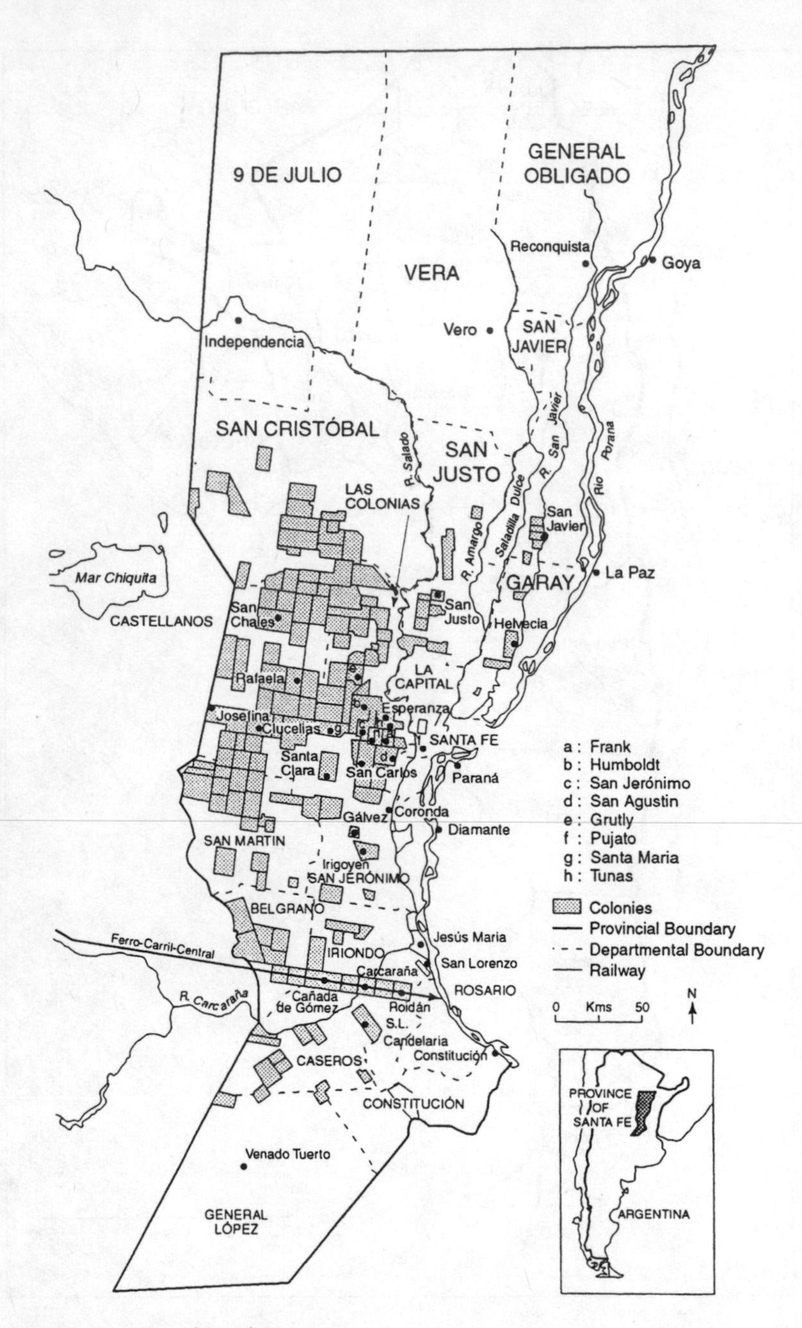

Map II Santa Fe and its colonias

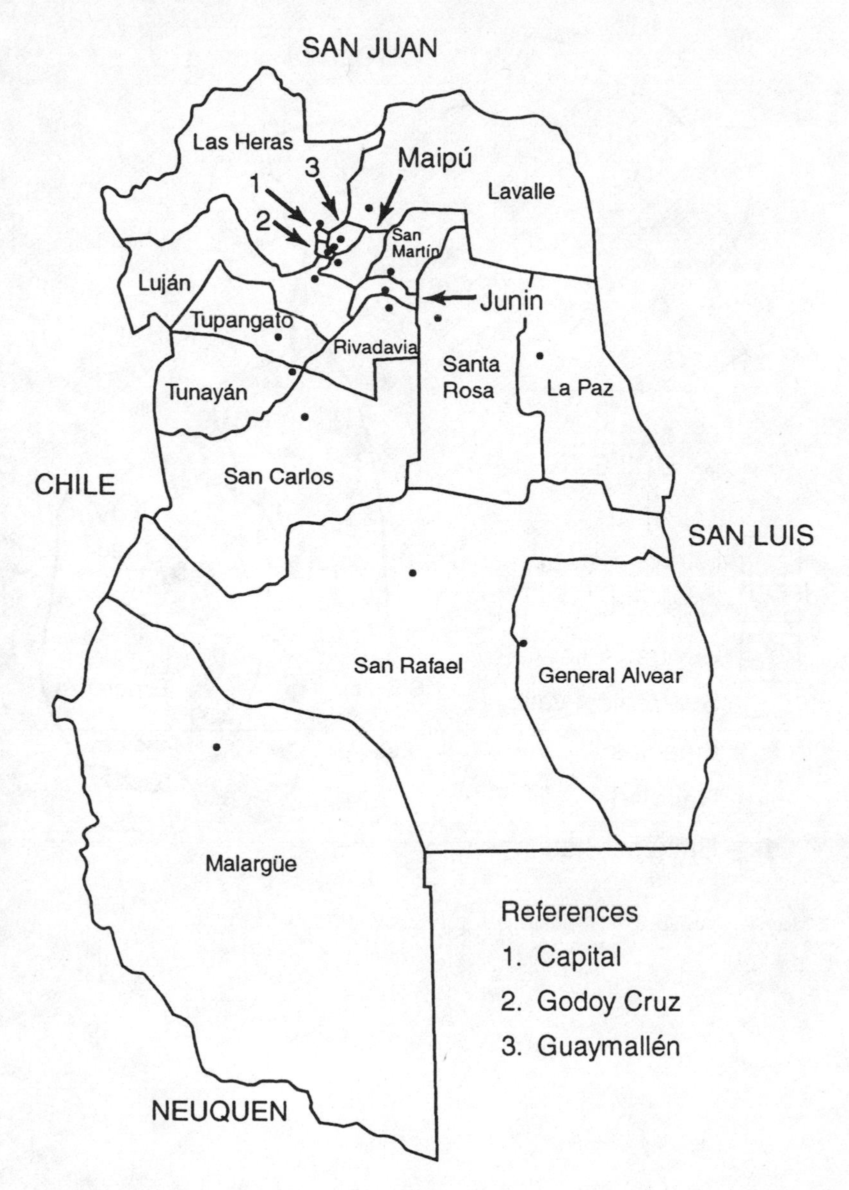

Map III Mendoza province and its political departments

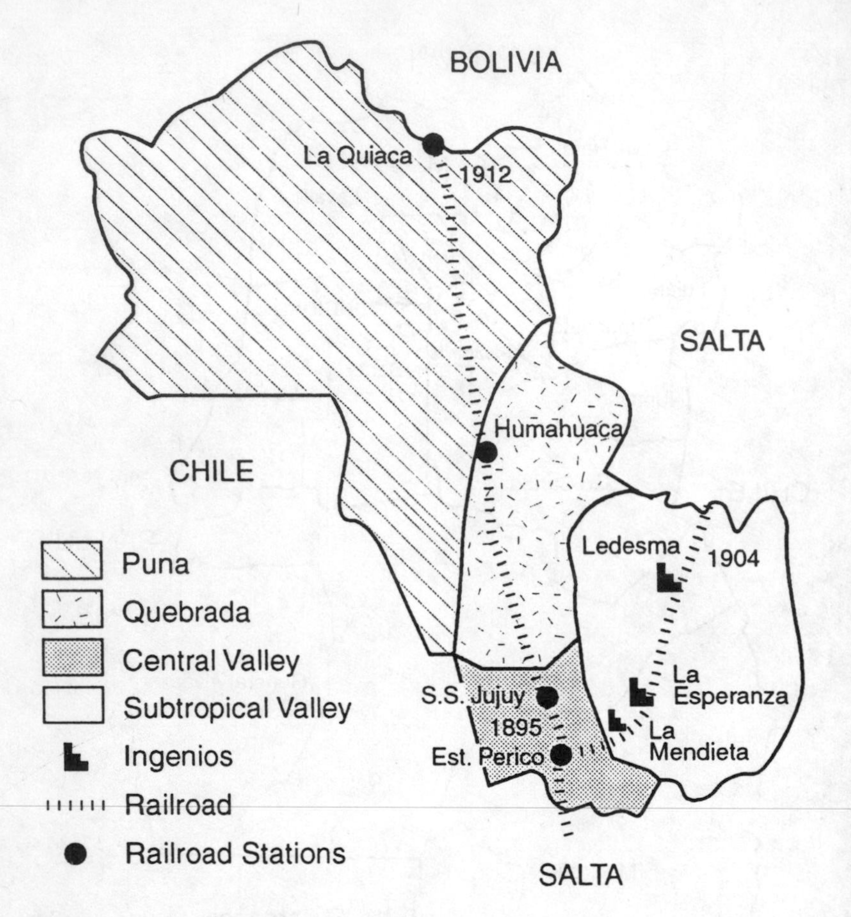

Map IV Province of Jujuy and its principal economic regions

Index